Vic Soffiantini
Analytical Chemistry

Also of Interest

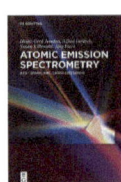

Atomic Emission Spectrometry.
AES – Spark, Arc, Laser Excitation
Device Physics, Fabrication, Simulation
Heinz-Gerd Joosten, Alfred Golloch, Jörg Flock, Susan Killewald, 2020
ISBN 978-3-11-052768-1, e-ISBN 978-3-11-052969-2

Micro-Raman Spectroscopy.
Theory and Application
Jürgen Popp, Thomas Mayerhöfer (Eds.), 2020
ISBN 978-3-11-051479-7, e-ISBN 978-3-11-051531-2

Elemental Analysis.
An Introduction to Modern Spectrometric Techniques
Gerhard Schlemmer, Lieve Balcaen, José Luis Todolí,
Michael W. Hinds, 2019
Together with: Publishing House of Electronics Industry
ISBN 978-3-11-050107-0, e-ISBN 978-3-11-050108-7

Elastic Light Scattering Spectrometry
Cheng Zhi Huang, Jian Ling, Jian Wang, 2019
Together with: China Science Publishing & Media Ltd.
ISBN 978-3-11-057310-7, e-ISBN 978-3-11-057313-8

Vic Soffiantini
Analytical Chemistry

Principles and Practice

DE GRUYTER

Author
Vic Soffiantini
4618 P.O. Box
Durban 4000
South Africa
ceo@laboratories.co.za

ISBN 978-3-11-072119-5
e-ISBN (PDF) 978-3-11-072120-1
e-ISBN (EPUB) 978-3-11-072121-8

Library of Congress Control Number: 2021947263

Bibliographic information published by the Deutsche Nationalbibliothek
The Deutsche Nationalbibliothek lists this publication in the Deutsche Nationalbibliografie;
detailed bibliographic data are available on the Internet at http://dnb.dnb.de.

© 2022 Walter de Gruyter GmbH, Berlin/Boston
Cover image: Vic Soffiantini
Typesetting: Integra Software Services Pvt. Ltd.
Printing and binding: CPI books GmbH, Leck

www.degruyter.com

Preface

This analytical chemistry textbook has been compiled based on a laboratory skills instruction manual format, to assist any laboratory worker who has not had the opportunities to obtain formal education and training in simple laboratory procedures or is reading undergraduate chemistry or a science course. It could also be used as a teaching aid for lecturers and professors, in general analytical principles and practice.

The book contains 25 chapters, many diagrams and photographs, as well as 20 appendices of useful tables and may be considered as an introduction to basic principles and operations or processes of analytical science, which is an applied science and art, whereby knowledge of the science, laboratory skills and the art, are used to obtain answers (e.g. for dissertations, laboratory data or test results) to problems such as:

- What test procedure to use for what type of sample composition or research work?
- What apparatus to use for that evaluation or investigation?
- How to use that apparatus or instrumentation?
- Is the test result or research data obtained precise, accurate and scientifically credible?
- Does the result show that the product, raw material, specimen, test or research item, or sample is within quality specifications and standards; or whether the results prove or disprove a hypothesis?

Keywords are listed at the beginning of each chapter, for ease of use and indexing.

There are some sections of text that has been repeated throughout the book. The intention of the repetition is to gain the attention of the reader or student to remember and note those sections or comments.

Note: whilst every reasonable care has been taken in compiling this technical chemistry book to ensure the accuracy of the information supplied, the author does not accept any responsibility for any errors or omissions or any consequences thereof.

https://doi.org/10.1515/9783110721201-202

Contents

Physical measurements

Introduction

Fig. intro: Chemistry and microbiology laboratory.

This book outlines analytical chemistry in a style and format that any student or new research scientist to laboratories and the practice of analytical chemistry can understand, without all the unnecessary scientific theory and mathematical equations. It can also be used as a reference book as there are 20 appendices containing helpful science jargons, definitions, critical information and useful scientific data.

Analytical chemistry is a crucial branch of general chemistry that undertakes the identification (**qualitative analysis**) of ionic species, chemical compounds and mixtures, as well as the determination of the levels or concentrations of these constituents (**quantitative analysis**), by using simple or very complex reactions and computerised instrumentation and software algorithms.

Analytical chemistry is a fascinating subject; it could be classed as an **art** as many technical operations require special **skills** and **talent** as well as creativity when problems occur or difficult analysis or separations of components of a substance are required.

The **chemical analyst** should know the **basic principles** of many other scientific and technical disciplines, such as mineralogy, microbiology, biochemistry, statistics, mathematics, software programming, computer science and IT networks, biotechnology, forensic science, photo-microscopy, chemical engineering, electronics and physics (fig. intro).

This makes the work of an analytical chemist a fascinating and thought-provoking career.

The **analytical chemist** is required to scientifically break things apart (analyse), inspect the parts (identify and measure), then put the whole thing together again (similar to reverse engineering) in a thesis or laboratory report.

https://doi.org/10.1515/9783110721201-001

A more representative and exciting career of an analytical chemist is that of an **analytical scientist** who incorporates many disciplines, such as those mentioned above, to arrive at answers to everyday problems.

In some areas of industry or commerce, the analytical chemist can be referred to as an **industrial chemist** or **public analyst** if working for a government agency; others are designations such as **forensic chemist** or **quality control chemist**.

Analytical chemistry is moving towards computerisation and artificial intelligence, where the general factory chemist is basically now just a keyboard operator or data logger. However, the basic principles of chemical analysis still remain the fundamentals upon which modern analytical and measuring instrumentation was built upon. The techniques of analytical chemistry and its computerised instrumentation are forever changing, presenting new challenges to the modern analytical chemist every day.

Today, the field and understanding of chemical analysis is expanding rapidly with the advent of advanced electronics, miniaturisation of instruments and computer algorithm calculations that took many hours (or days) of labour in the past, which now takes only several minutes before test results can be submitted worldwide over the Internet in seconds.

Analytical instrumentation with automatic samplers and measuring instrumentation is making the life of the analyst easier, but with more challenges, such as **reducing the detection limits**, **improving precision** of procedures, developing **new methodology** for the many new compounds and products that are being manufactured daily.

A distinction can be made between that of **analytical chemistry** and **chemical analysis**, by inferring that analytical chemists (analytical scientists) work to improve and extend established test methods, whereas the chemical analyst makes use of laboratory skills and current knowledge to undertake the daily tasks of a chemist.

Chapter 1
Analytical chemistry

Keywords: qualitative, quantitative, electromagnetic radiation, spectroscopy, optics, rheometry, chromatography, thermodynamics, electrochemistry, physical methods, hybrid techniques, scale of analysis, sensitivity levels, identification

Fig. 1.1: Chemical tests.

This chapter outlines what analytical chemistry entails for the student, research worker and industrial quality control laboratory staff, as well as information for technical and scientific consultants.

What is analytical chemistry?

It is an applied field of expertise, where basically a procedure of breaking down (**extracting**) the component parts of a substance or item, **identifying** those parts and configuring how they all **fit together.**

In order to do this, the analytical chemist needs resources, such as laboratory testing procedures, standard operating procedures and policies, measuring equipment and recording instrumentation as well as computer software programmes to analyse the output data (Fig. 1.1).

Analytical chemistry consists of essentially **qualitative analysis** (what is present in a substance) and **quantitative analysis** (how much is present in the substance). By definition, qualitative analysis does not measure the mass or concentration of an analyte. Quantitative analysis, by definition, measures the mass or concentration of the analyte sought.

https://doi.org/10.1515/9783110721201-002

Inorganic qualitative analysis generally refers to **elemental** analysis by using a systematic procedure of elimination, using techniques of separation and identification. This is sometimes referred to as microchemistry as these procedures are undertaken on trace quantities of the unknown substance.

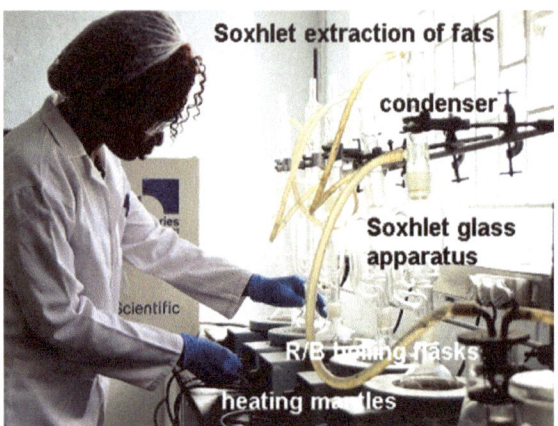

Fig. 1.2: Soxhlet fat extraction.

Analytical chemistry may be broken down into several classes or patterns of analytical techniques and various scientific procedures. These can be generalised below, based on their **chemical**, **physical** and **electrical** principles of basic operation:
- **Electromagnetic radiation (spectroscopy)**; e.g. inductively coupled plasma (ICP), atomic absorption spectroscopy (AAS), ultraviolet (UV) spectroscopy, infrared spectrometry, nuclear magnetic resonance (NMR), mass spectrometry (MS), X-ray diffraction.
- **Optics**, e.g. refractometry, polarimetry, nephelometry and turbidimetry.
- **Rheometry**, e.g. viscosity.
- **Chromatography** and **ion exchange**, e.g. gas chromatography (GC), high-performance liquid chromatography (HPLC), ion chromatography, gel electrophoresis.
- **Thermodynamics**, e.g. calorimetry, thermal analysis.
- **Electrochemistry**, e.g. conductometry, potentiometry, pH, ion-selective electrode, voltammetry, coulometry, electrodeposition, amperometric titrations.
- **Classical wet chemistry**, basically these techniques consist of **qualitative** analysis involving spot tests, microchemistry, flame tests and **quantitative** analysis consisting of **gravimetric** (mass measurements before and after specific chemical reactions) and **volumetric** analysis (stoichiometry).
- **Physical methods**, e.g. density and particle size distribution.

– **Hybrid techniques**, e.g. GC-MS, ICP-MS, liquid chromatography-NMR; where the former instrument undertakes the **separation** process and the latter instrument does the **identification** and measurement (usually acts as an extra detector).

A **determination** is considered to be the measurement of a single constituent (**analyte**) of a sample or item.

An **analysis** is considered to be the **preparation of the sample** for the test parameters required, the **selection** of an appropriate test procedure and the **measurement** of the levels of constituent parts of the sample or item.

The **scale of the analysis** is determined by the amount of sample available and the expected level of concentration of the analyte (also depends on laboratory equipment available).

As a guideline to **sensitivity levels** of test procedures and a representative amount of the sample, use the following scale:
– major analyte constituents in % (g) levels (i.e. **macroanalysis**), require at least 1 g sample for analysis;
– minor analyte constituents in 0.01% (centigram) levels (i.e. **semi-microanalysis**), require more than 2 g sample;
– minor analyte constituents in 0.001% (mg) levels (i.e. **microanalysis**), require 5 g sample;
– trace analyte constituents (microgram) levels (i.e. **ultra-microanalysis**), require at least 10 g sample for accurate and representative analysis.

What this basically means is that, at least 10 g sample is required to detect ppm or microgram levels of an analyte, using general analytical (aka wet chemistry) techniques, such as gravimetric and volumetric test methods; otherwise use instrumental analysis such as GC, HPLC, MS and Fourier-transform infrared (FTIR), which require only a few milligrams sample.

Modern analytical chemistry is dominated by highly intricate computer controlled expensive instrumentation, requiring highly experienced analysts as operators, decision makers and problem solvers. Research is continuously undertaken to improve on the selectivity, sensitivity and robustness of analytical techniques and instrumentation.

The **identification** of unknown substances (organic or inorganic compounds) can be done by using the technique of **fingerprinting**. This is basically comparing a pattern (e.g. spectrum) of the unknown to that of a pattern of a known substance. This can be done by using sophisticated instrumentation, such as FTIR spectrophotometry, X-ray diffraction and GC-MS.

Chapter 2
Laboratory operations

Keywords: tools of lab, solvents, thermometers, temperature, filtration, lab procedures, lab criteria, ISO 17025, types of glassware, heating devices, extraction solvents, filter paper

Fig. 2.1: Laboratory testing.

Many aspects of how procedures are undertaken in laboratories are discussed in the following paragraphs.

An outline of the functions of a laboratory, its resources, policies and operations and its purpose within an organisation is given here.

A **laboratory** is a place of work that undertakes various testing tasks (an example is microscopic examination of forensic items, as shown in Fig. 2.2 below), or experimentation and reports (publishes) the results (data) obtained.

The laboratory is thus a service department which **generates** information and data from its test results or experimentation (e.g. quality control or research data), and **submits** these to another department or persons (e.g. marketing division, factory manager, thesis supervisor, director of R&D).

The aim of the laboratory analyst is to obtain the most conclusive scientifically and statistically credible answer to chemical and scientific requests, in the shortest length of time, using whatever is at the disposal of that laboratory.

The laboratory must ensure that the information and/or data that it generates is:
- as accurate and as precise as possible (Fig. 2.1);
- obtained from calibrated and verified measuring instrumentation and apparatus;
- only validated test methodology is used, irrespective of whether it is routine test procedures or new methodology;

https://doi.org/10.1515/9783110721201-003

- laboratory reports on time, i.e. no unnecessary delays in the submission of results or data;
- supplied to the correct departments, professors or other interested persons;
- in the correct format, i.e. reported on the right form, spreadsheet, computer programme, graph or chart; and
- reported by highlighting any deviations from any given specifications or experimentation (or company control limits), and making any recommendations or a null hypothesis, if required.

In order to meet the above criteria, the laboratory must:
- be competently organised, having available written standard test procedures and manuals that can be understood by all persons concerned; for example, a **Quality Laboratory Manual** such as the **ISO 17025**; supplemented by in-company or university policy manuals (standard operating procedures, aka SOP) or international test procedures (e.g. ASTM, AOAC, AOCS, EN or ISO standards, USP or BP monographs);
- be **clean and tidy**, with a place for everything and everything in its place; cleanliness is essential in order to avoid, amongst other things, contamination of specimens, samples; or the soiling, breakdown of computerised equipment and other sensitive measuring instrumentation;
- be a safe and harmonious working environment for all laboratory workers, that is conducive to **good laboratory practices**;
- must comply with government regulations, such as the Occupational Health and Safety Act;
- **employ** appropriately trained and experienced laboratory staff and supervisors to undertake the assigned duties;
- employ a suitably science-qualified, trained and experienced person in charge to undertake the supervisory duties of the daily administration, stock taking of chemical reagents, co-ordination and control of the laboratory department;
- keep **records** (preferably on Cloud programmes or movable hard drives) of all test results, data and reports for at least 5 years, so that any queries (e.g. customer complaints, further research) or trends (e.g. shelf-life of products) may be processed or evaluated;
- have all **equipment** and instrumentation checked, **calibrated** and logged regularly, as per the organisation or university SOP;
- **regularly review** the programmes for frequency of sampling, tests to be done and test methods to be used, as per the organisation's SOP;
- be up to date with the latest trends in analytical research and instrumentation.

Thus, one can say that, raw materials and products (i.e. test items, specimens, samples, vials and swabs) **flow into** a factory or research laboratory and information (e.g. test results, reports, Certificates of Compliance, dissertations) **flows out** of the laboratory.

This all takes place in a laboratory.

Simple laboratory procedures:

– taking temperature measurements, such as before sampling of liquids, or noting temperature of water baths or incubators used for conducting tests at specific temperatures, such as viscosity and density measurements, or incubation of bacterial flora;

– measuring the density or specific gravity of a liquid using hydrometers, automatic density meters or pycnometers, to establish the strength or purity of a liquid;

– measuring the refractive index of a liquid using a refractometer to establish the strength, purity or identity of a liquid (or solid or gas);

– measuring the turbidity or colour of a liquid using a turbidity meter or colorimeter to establish the strength or purity of the liquid; or even establishing the bacterial load of a mother culture (haemocytometer); here, spectrophotometers can also be used for measuring turbidity and for counting bacterial cells;

– measuring or dispensing volumes of liquid samples or other liquid chemical reagents using automatic syringes, measuring cylinders, pipettes and burettes (titrimetric/volumetric analyses);

– removing solid matter or precipitates (gravimetric analyses) from liquids using filtration procedures by vacuum, gravity or pressure; an alternative is by centrifugation;

– measuring the electrical conductivity of liquids and hence calculating, by using a predetermined conversion factor, the total dissolved solids (TDS) and salinity of waters and effluents;

– measuring the acidity or alkalinity of substances using indicator papers, test kits, colour indicator solutions, natural dyes (such as litmus) or pH meters and probes;

– weighing (aka massing out) known amounts of sample or chemicals using balances (mass meters);

– titrations based on principle of stoichiometry (aka volumetric work), e.g. acidity test on milk, testing for sodium chloride in seawater or brine;

– gravimetric work, e.g. precipitations, filtering and weighing (moisture testing);

– determining the moisture content of pastes, sludges or solids by oven drying at various temperatures depending upon the nature of the solid or item, also at various controlled humidity levels. There are other various procedures for determining moisture content or the water of crystallisation, such as Karl Fischer titrations, infrared drying or even by acetylene gas pressure measurement, by reaction of calcium carbide with moisture (the carbide method).

– Using simple colour comparisons or measurements; examples are the Maxwell colour charts and Lovibond (tintometer) comparator colour discs.

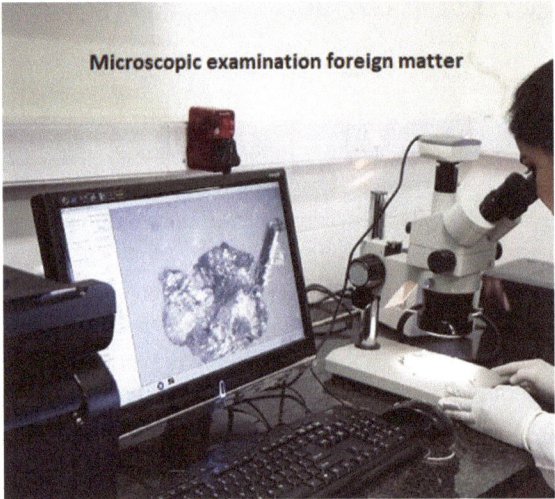

Fig. 2.2: Microscopic examination.

Tools of a laboratory worker:

The basic tools and equipment used in most industrial and university chemistry laboratories are listed in Tab. 2.1 and 2.2 below:

Tab. 2.1: Tools.

Equipment	Uses	Examples of sample types
pH meter	Measures degree of acidity or alkalinity	pH of water; acidity of fruit juices
Ion-selective electrode meter (ISE)	Measures cations and anions, but best for fluorides (and pH)	Fluorides in drinking water and toothpaste
Nephelometer or turbidimeter	Measures amount of suspended matter or degree of turbidity	Turbidity of effluents; stability of emulsions
Conductivity meter (preferably a model with automatic temperature compensation, adjustable cell/probe constant (TDS factor or K factor)	Measures electrical conductivity/resistivity of ions	Purity of water; conductance of ions; TDS of boiler waters, salinity of salt water or brine; checks for static potential in jet fuels

Tab. 2.1 (continued)

Equipment	Uses	Examples of sample types
Dissolved oxygen meter (preferably a model with automatic temperature compensation, automatic salinity correction, barometric pressure correction)	Measures oxygen gas levels in liquids	Dissolved oxygen in waters, effluents and sewage ponds; fish breeding tanks
Flame (emission) photometer	Detects Li, Na, K, Ca, Ba and Mg	Salt analysis, fertilisers, foodstuffs, waters
Atomic absorption spectrophotometer (AAS)	Detects most metals at trace levels. Measures absorbed energy of excited ions	Metals (cations) in food analysis, water; pharmaceuticals, fertilizers, heavy metals in food and cosmetic products
ICP-OES	Similar to AAS but measures more than one element/metal at a time. Measures emitted energy of ionised ions (plasma gas)	Heavy metals and a few non-metals
Polarograph	Detects different species (e.g. different oxidation states of metals or functional groups of organic compounds)	Fe(II) and Fe(III) separately; organic species; pharmaceuticals
Colorimeters	Measure's intensity of visible colour in solutions (and solids)	Phosphates in waters, food, fertilizers, dyes
UV-visible spectrophotometer	Measures intensity of colour in solutions, from about 200 nm to 850 nm wavelength	Phosphates, pharmaceuticals, solvents, waters

Note that an **ion-selective electrode meter** can also be used as a **pH meter** or an oxidation–reduction potential meter.

In the laboratory, the chemist will use various types of glassware, such as those illustrated below:

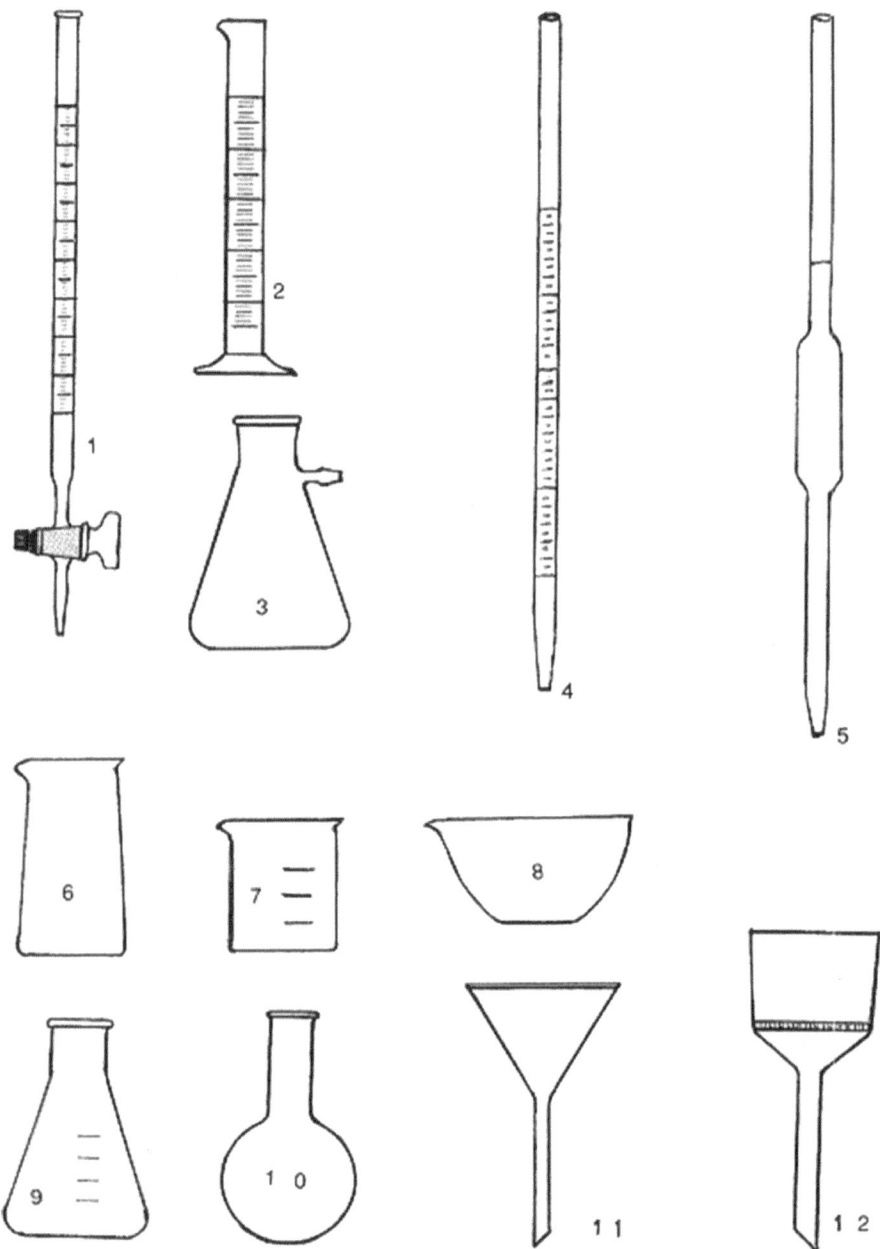

Fig. 2.3: Types of glassware.

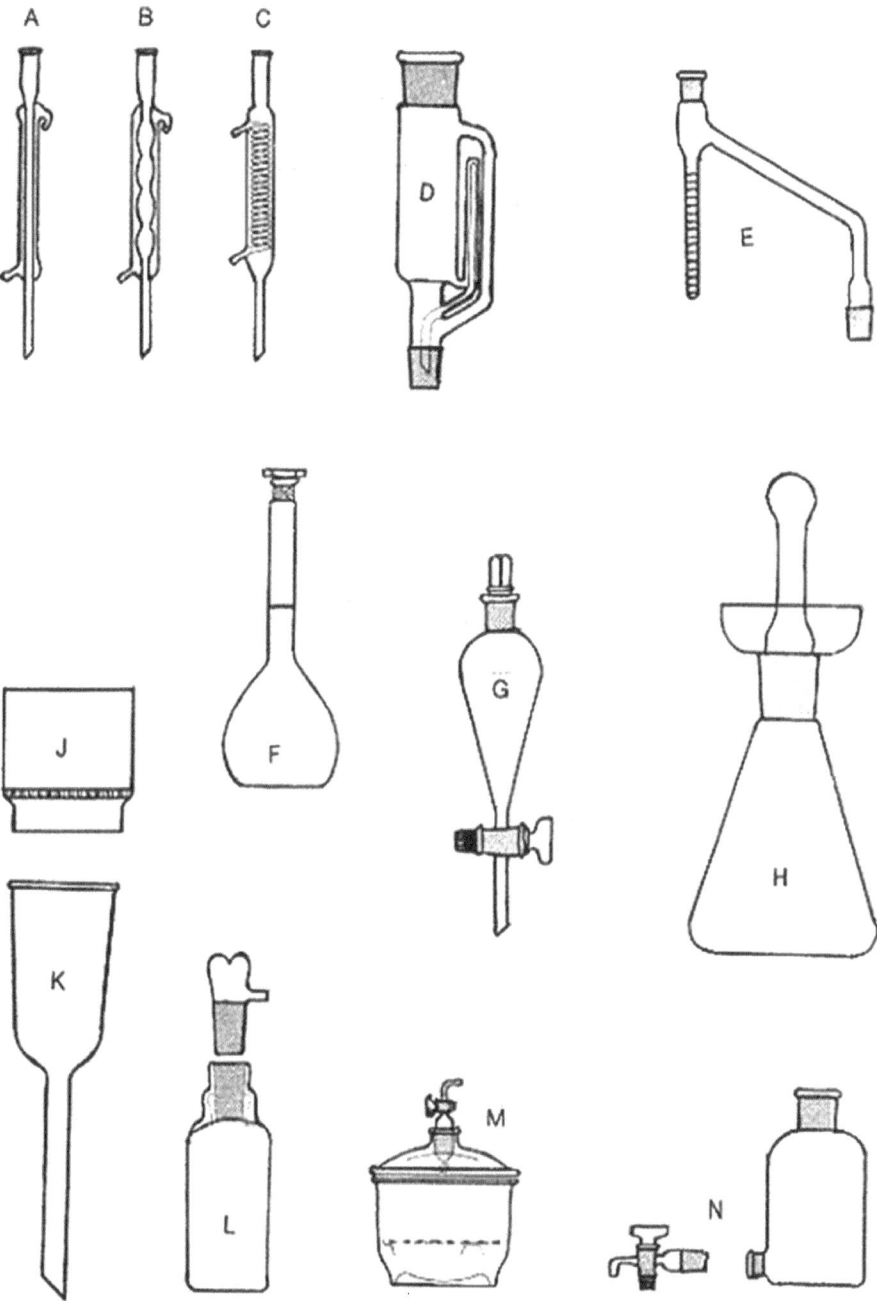

Fig. 2.4: Types of glassware.

Legend to Figs. 2.3 and 2.4:

Figure	Name of glassware	Examples of uses
1.	Burette	Titrating or dispensing accurate and precise volumes of liquids.
2.	Measuring cylinder	Dispensing approximate volumes of liquid.
3.	Vacuum flask (aka filtering flask)	Filtration used in combination with J&K below.
4.	Graduated pipette	Dispensing accurate and precise increment volumes of liquid.
5.	Bulb pipette	Dispensing a certain volume of liquid accurately.
6.	Phillips beaker	Container suitable of holding hot liquids or for general use.
7.	Beaker	General-purpose container.
8.	Evaporating dish	Evaporating or boiling down liquids under heat.
9.	Conical or Erlenmeyer flask	Container used in titrations or for general-purpose use.
10.	Round bottom boiling flask	Container used for boiling liquids in heating mantles.
11.	Funnel	Pouring liquids into narrow neck containers.
12.	Buchner funnel or filter	Filtering liquids and so on under vacuum.
A.	Liebig condenser	General-purpose condenser for distillation and refluxing.
B.	Allihm condenser	More efficient refluxing condenser (greater surface area) than the Liebig condenser, e.g. for low boiling solvents, ether.
C.	Graham condenser	Very efficient condenser for refluxing liquids.
D.	Soxhlet extractor	For extraction of solids, e.g. fat determination using solvents.
E.	Dean and Starke receiver	Separation (by refluxing) of liquids, e.g. water determination in oils and feeds, using toluene or xylene as an azeotropic mixture.
F.	Volumetric flask	Container for a fixed (accurate at designated temperature) volume of liquids.
G.	Separating funnel	Separating two or more liquid phases.
H.	Iodine flask	Special titration vessel for iodimetric titrations.
J.	Sintered glass filter crucible	For separation and filtration of liquids, under vacuum (see 3 above).
K.	Holder for filter crucible	Holder for filter crucibles under vacuum (see 3 above).

(continued)

Figure	Name of glassware	Examples of uses
L.	Dropping bottle or indicator bottle	For dispensing drops of liquid, e.g. indicators
M.	Desiccators (vacuum) and glass	For drying or storing items under specific humidity conditions.
N.	Aspirator bottle	Storage and dispensing container via a tap.

Tab. 2.2: Heating devices.

Type	Description	Uses	Temperature range
Steam baths	Generally, a stainless steel unit with concentric rings. Rings are removed according to the size of the object (e.g. beaker). Can take round bottom flasks and flat base beakers. The unit has water inlet adapter as well as an overflow outlet. Water is used to fill the bath and to keep bath full water via running water through the inlet adapter. Some units have a thermostat heating controller for the immersed element. The steam does the heating of the objects placed on lid but sometimes a rack of test tubes can be immersed in the boiling water if test method states such heating required.	To evaporate down volatile liquids, especially flammable solvents; also used to digest or dry precipitates. Can be used for incubation of cell cultures.	Maximum 100 °C at sea level.

Tab. 2.2 (continued)

Type	Description	Uses	Temperature range
Water baths	An insulated container containing water and fitted with a thermostat controller and stirrer unit. The objects are immersed in the water, but precautions are taken not to allow bottles to "float" or tilt over; or allow water ingress into the sample bottles.	To heat objects at various precisely controlled temperatures via thermostat. Used in chemical reactions or to incubate bottles, microbiological media, or solids in liquid form until needed for use. Evaporate off volatile compounds. Ideal for maintaining constant temperature of liquids and solids.	From room temperature to just below boiling point of the water.
Metal baths	An insulated container containing special reusable thermal metallic alloy beads (aka shot) in place of water; fitted with a thermostat and temperature controller. Advantage is that this type requires less energy than heating water and does not require regular cleaning like water baths. Objects are placed into the bed of (aluminium) beads.	Disadvantage is that it takes a long time for temperature uniformity. Ideal for keeping items hot (e.g. autoclaved media or agar) Without having to top up water levels; also, items do not "float" if not weighted down which is a constant problem with water baths. Evaporate off moisture, water or high-temperature solvents.	Room temperature to about 350 °C depending upon type metal beads
Sand baths	Same as water bath and metal bath but contains sand instead. The sand bath is usually placed on a hotplate as its source of heat. The objects are imbedded in the sand or may be placed on top of the layer of sand; this depends upon the size and shape of the object/beaker/flask. The sand Used is generally fine silicate-type sand such as acid-washed beach sand.	Used for chemical reactions, ashing, drying or even boiling liquids. Also used for evaporating flammable solvents. Heating is done by conduction. Suitable for heating substances or boiling at high temperatures. If acid boils over in the flask, then just change the contaminated sand; no damage done to hotplate underneath sand bath.	Room temperature to about 350 °C

Tab. 2.2 (continued)

Type	Description	Uses	Temperature range
Oil baths	Bath is similar to a general-purpose water bath but contains a high boiling point liquid in place of the water. Oil used is generally a clear colourless mineral or silicone oil of a high smoke point.	Ideal for use as a viscosity bath. Generally used when temperatures greater than 100 °C are required. Heating is done by conduction. Must ensure that no water or wet objects are placed in these baths.	Room temperature to about 200 °C or higher depending upon temperature degradation of the bath oil. Ideal for uniform heating at high temperatures.
Hotplates	Can be a simple hotplate or a combined hotplate stirrer. Has a thermostat and a temperature controller	Must ensure that no wet objects are placed on a **hot** hotplate, especially glassware! Not suitable where flammable vapours or solvents are present! Can be used for charring organic samples. Generally used for heating and stirring liquids.	Room temperature to about 450 °C
Heating mantles	Disadvantage is that liable to acid or mechanical damage of the wire element imbedded in heat-resistant cloth. Aka isomantle.	Suitable for round bottom distilling/refluxing flasks. No "hotspots" present unlike hotplates that tend to have uneven heating of the plate surface. Ideal for refluxing with condensers or Dean and Stark apparatus.	70 °C to about 250 °C
Gas burners	Propane/butane, Handigas, Cadac; butane/pentane. Bunsen burners, Meker burners. Meker–Fisher burners	Not suitable where flammable vapours or solvents are present! Low-temperature flame can be used for sterilizing small articles such as loops and forceps (tweezers). Ashing or burning off carbon from organic substances/samples.	Depends on air/gas ratio settings, usually 3:1. **Blue** flame is a hot (±2,000 °C) **oxidizing** flame; **yellow** flame (±1,000 °C) is a luminous/safety/**reducing** flame.

Tab. 2.2 (continued)

Type	Description	Uses	Temperature range
Ovens	Lab ovens should be explosion proof. Heating is by convection; but some ovens are equipped with fans, i.e. forced ventilation.	Do not dry or place objects in ovens that could cause an explosive mixture. Heating is done by convection. Do not place articles that would melt or deform due to no safety setting of the temperature of the oven. Most common use is for moisture determinations.	Most common range is from 50 to 250 °C, but some models go as high as 330 °C.
Vacuum ovens	Various shapes, sizes and wattage range	Used generally for drying heat-sensitive materials, where low temperature under a vacuum increases the rate of volatility of water or other solvents. Desiccating storage of air sensitive materials. Drying of non-flammable volatile substances.	Depends upon vacuum gauge setting and oven thermostat setting. Note that **temperature does not change under vacuum** since its energy cannot move (no convection). Temperature will be as is, in the oven irrespective of degree of vacuum applied.
Incubators	The use of the extra glass door is to be able to observe what is happening inside the incubator without losing any heat by opening the door; also, a second door assists to control constant temperature inside.	Primary use is for growing bacteria or cultures at a specific and precise temperature, for specified time periods; accurate control of temperature is critical.	Generally, from ambient temperature to maximum 70 °C
Furnaces	There are various types, such as Muffle furnaces and tube furnaces. Forced ventilation or slightly ajar door, to produce an oxidizing atmosphere; otherwise closed system to produce a reducing atmosphere.	Tube furnaces are used for measuring the Ash Deformations of coal sample. Also, special "volatile matter" furnaces are of special dimensions to comply with ASTM specifications. Ignition or ashing of samples.	Depending upon wattage, ranges are 1,000 °C, 1,200 °C and 1,400 °C.

Tab. 2.2 (continued)

Type	Description	Uses	Temperature range
Heating tapes	There are various lengths and temperature ranges, depending upon specific uses.	A useful use is to prevent the freezing of a nitrous gas regulator valve, when used in AAS.	Maximum usually 200 °C
(Dry) heating blocks	Various sizes and temperature ranges from cooling to heating.	Commonly used for small chemical reactions or spot test kits. Ideal for mini-COD tests or digestion of sample volumes of substances.	Maximum usually 250 °C
Autoclaves	Various sizes. Sterilization can be "dry" or "steam"	Used for sterilizing objects, bottles, glass petri dishes and microbiology media and broths. Autoclave tape or live cultures are used to determine effectiveness of sterilization.	Generally, 121 °C at 15 psi
Microwave ovens	This must be explosion proof and acid resistant, not the domestic type.	Heating is by radiation. Items must not contain any aluminium because it could cause explosions.	Depends on wattage rating

Solvents:

A **solvent** is usually a liquid in which a **solute** (usually a solid substance) is dissolved to form a **solution** (usually a composite liquid). Refer to Tab. 2.3 below for examples of different types solvents for the extraction or dissolution of different types materials.

The quantity of solute that can be dissolved in a specific volume of solvent depends upon temperature and other interacting forces. Note that e.g. 100 mL alcohol to which 100 mL water has been added does not produce a total volume of 200 mL but less, due to intermolecular forces.

There are, by convention, **two basic types** (depending upon their strength in promoting ionisation of the solute) solvents in analytical chemistry; namely:

Ionising solvents, such as water, hydrochloric acid, acetic acid, ammonia, sulphur dioxide and amines; these can be further divided into **protonic solvents** (e.g. water and mineral acids) and non-protonic solvents (e.g. sulphur dioxide);

Non-ionising solvents such as chloroform, hexane, benzene, ethers and esters.

Saturation is where no further amount of solute can be dissolved in a fixed volume of solvent, at a specified temperature and pressure.

The **solubility product constant**, Ksp, of a chemical or solid, is the **equilibrium constant** at which a solute dissolves into solution, to form a saturated solution.

$$K_{sp} = [A^+]a \times [B^-]b$$

where A^+ is the cation in an aqueous solution, B^- is the anion in an aqueous solution
a, b are relative molar concentrations of a and b.

The **common-ion effect** is a shift in chemical equilibrium which affects the solubility of solutes in a chemical reaction. This is sometimes referred to as the **Le-Chatelier's principle** of equilibrium reactions. This principle states that when a system in dynamic equilibrium is acted upon by an external stress, the system will change its equilibrium status in order to relieve the stress exerted upon it.

Basically, the common-ion effect occurs when there is same ion present in both a reacting compound and a product compound, causing a suppression or lowering of the solubility of one of the compounds.

An example of this effect is in the **buffering** status or equilibrium pH status of salts, such as the reaction between ammonium hydroxide (NH_4OH) and ammonium chloride (NH_4Cl), whereby the ammonium chloride suppresses the ionisation of ammonium hydroxide causing a decrease in hydroxide ion (OH^-) concentration and hence increase in hydrogen ion (H^+) concentration, thus changing the pH status of the solution leading to a more acidic (less alkaline) pH level. Note that the common ion here is the NH_4^-.

Other examples are the dissociation of weak acids (and weak bases).

An example in **gravimetric analysis** is where the precipitation of any electrolyte (compound) occurs when the concentration of its ions is increased by addition of a common ion; thus exceeding the **solubility product** of the electrolyte (compound).

Tab. 2.3: Extraction solvents.

Sample type	Solvent	Flash point (cc) in °C
Vegetable oils	n-Hexane (hexyl hydride)	−22
Oleoresins	Methylene chloride (dichloromethane)	Non-flammable
Spices	Methylene chloride (dichloromethane)	Non-flammable
Decaffeinated beverages	Water or methylene chloride	Non-flammable
Hops and malt	Ethyl alcohol (alcohol) (ethanol)	12
Polystyrene	Methyl isobutyl ketone (MIBK) (hexone)	14
Cellulose acetate	Methyl isobutyl ketone (MIBK) (hexone)	14
Cellulose acetate-butyrate	Methyl isobutyl ketone (MIBK) (hexone)	14
Polyacrylonitriles	Dimethylformamide (n-formyldimethylamine)	58
Polycarbonates	Dimethylformamide (n-formyldimethylamine)	58

Tab. 2.3 (continued)

Sample type	Solvent	Flash point (cc) in °C
PVC and PVA copolymers	Cyclohexanone (cyclohexyl ketone), dimethylformamide	44
Polyamides	60% formic acid	–
Polyethers	Methyl alcohol (methanol)	9
Perspex	Dichloromethane	Non-flammable
uPVC	Methyl ethyl ketone (2-butanone)	–9
Rubbers, natural	Cyclohexane (hexamethylene)	–18
Bitumen	Benzene fraction	–10
Plastics and varnishes	Acetone (dimethyl ketone)	–18

Note: The general idea is that "like dissolves like"; which means **polar solvents** are good for dissolving **polar compounds** and **non-polar solv**ents readily dissolve **non-polar compounds**. Polar liquids are generally immiscible with non-polar liquids.

Some possible extraction gases are:
– carbon dioxide;
– nitrous oxide (laughing gas! – anaesthetic!);
– ethylene;
– chlorotrifluromethane;
– trifluoromethane.

Thermometers and temperature:
Temperature is a degree of heat (or cold) where energy moves from matter of higher temperature to matter of lower temperature, until temperature equilibrium is reached. This heat transfer process is by convection, conduction or radiation.

A **thermometer** (aka temperature sensor) is a device capable of measuring the temperature of a substance.

There are basically five types of thermometers:
– **dial** thermometers, e.g. oven or fridge thermometers;
– **glass** mercury- or alcohol (spirit)-filled thermometers, e.g. laboratory thermometers or the obsolete clinical thermometers.

There are basically three types of glass thermometers:
– partial immersion to 76 mm, or to 0 °C (ring around thermometer);
– partial immersion up to mercury column; and
– total immersion of whole thermometer.

The sale and use of mercury type have been discontinued in many countries, due to the toxicity of mercury. However, these types are the most accurate and have very quick temperature response with respect to rate of expansion or contraction of the medium.

- **digital thermometers**, e.g. portable battery-operated meters that cover a wide temperature range depending upon the type of probe (e.g. Pt100 for readings below or at 100 °C; or thermocouples for readings above 100 °C);
- **Infrared**-type temperature measuring instruments; there are various temperature ranges, e.g. those used for measuring very high temperatures (such as 1,500 °C) where access is difficult or impossible, e.g. furnaces. And those used for low temperatures, e.g. baby thermometers and for monitoring virus or flu infections.
- **Resistance temperature detectors** or probes, such as thermistors and thermocouples (thermopiles). These are sometimes used in temperature recorders or data loggers.

Generally:

Dial thermometers are not accurate, but are easy to display and are robust if handled carelessly; these rely on the thermal expansion of metals; accuracy about 10%;

Digital thermometers are used extensively in various fields of industry and medicine. Good-quality electronic ones are used in various instruments; however, their accuracy (not precision) is very much dependent upon calibrations and the type and quality of the measuring probe. Digital thermometers often use thermocouples as the sensors. Accuracy could be 0.1% but is usually about 1% depending upon the quality of the sensing probe.

Indicating and recording (chart) thermometers are used in factories, e.g. time chart or data logging recorders. Accuracy is about 5%.

Galileo thermometer is an ornamental type transparent cylinder of water containing coloured glass bubbles of different densities indicating an approximate temperature of the room.

The analytical chemist must decide which type of thermometer or sensor would be suitable for the type or nature of the experiment or test procedures to be undertaken.

Most commonly used analytical or laboratory instrumentation have attached or imbedded types of thermistors or thermocouples. Otherwise, use is still made of calibrated glass thermometers which are the most accurate and are used as reference thermometers.

A **thermistor** is actually a resistance thermometer (aka **thermostats**) which senses temperature changes via its change in resistance which is measured by changes in electrical voltage charges (pd) of the metallic probe.

There are two types of thermistors, namely **negative temperature coefficient** (NTC) and **positive temperature coefficient** (PTC).

The NTC is where the resistance decreases when the temperature increases.

The PTC is where the resistance increases when the temperature increases (e.g. fuses). This type is often used in laboratories and is known as the Pt100; where Pt stands for platinum (the metallic probe) and has a resistance value of 100 ohm at 0 °C.

Thermistors are precise measuring sensors and usually their outputs are accurate to about 1/10th of a degree or better depending upon the electrical circuitry and construction of the apparatus.

A thermocouple is basically a pair of two dissimilar conducting wires bonded at one end (the sensing end) to form a junction and the other end of the two wires connected to a voltage (pd) measuring electrical circuit (the thermoelectric or **Seebeck effect**).

There are basically four types of thermocouples depending upon the temperature range to be measured or monitored.
- The J type (iron–constantan material) has a range of about −190 °C to +700 °C; i.e. for low temperatures.
- The K type (chromel–alumel material) has a range of about −250 °C to +1,150 °C; i.e. for high temperatures.
- The T type (copper–nickel material) has a range of about −260 °C to +860 °C; i.e. for low temperatures and is more resistant to corrosion than J type.
- The E type (nickel–chromium material) has a range of about −270 °C to +800 °C; has better accuracy than the K type or J type.

There are many other types used in industry, such as B, R and S types for very high temperatures.

Thermocouples are not precise measuring sensors and usually their outputs are not better than the tolerance of 2 °C, depending upon the type and temperature range.

Important: All thermometers must be regularly (at least annually) **calibrated** against a certified **reference thermometer**. These calibrations/checks must be done, at least, on three different temperature points, preferably at those temperatures most commonly measured in the laboratory.

Filtration

The **principle** here is a process of **separating** a **suspended solid**, such as a precipitate, from the liquid, by straining it through a porous medium that can be penetrated easily by the liquids; but not the solid particles (e.g. the precipitate).

Filtration is a basic manufacturing process in the chemical manufacturing industry and is also employed for such diverse purposes as the brewing of coffee, the clarification of sugar syrups or juices and the treatment of sewage. The liquid to be filtered is called the **suspension**. Filtration is also a very common process in laboratory work. The liquid that has passed through the filter is called the **filtrate** and the solid material remaining on the filter is known as the **residue** (or filter cake).

Some types of laboratory filtration mediums are:
- Filter paper (selected whether by gravity or by vacuum filtration; filtration by applying air pressure can also be done where gravity suction is inefficient)

- Sintered-glass filtering crucible (use vacuum filtration). Refer to Tab. 2.4 below for the sizes of filter crucibles.
- Gooch crucibles with glass fibre (asbestos is no longer used due to its hazardous nature) mat (use vacuum filtration)
- Glass fibre filters;
- Buchner funnels with filter paper by vacuum filtration
- Extraction thimbles
- Membrane filters (sterile or non-sterile);
- In-line filters
- Syringe filters

Maximum temperatures of filtering crucibles for ignition of precipitates or residues:
- Glass crucible = 450 °C
- Porcelain crucible = 1,000 °C
- Aluminium oxide crucible = 1,300 °C

Note that sometimes a **filter aid** is necessary when filtering fine or gelatinous suspensions; aids such as **diatomaceous earth** (e.g. Kieselguhr, Perlite or Celite) are generally used. These substances are unreactive silicate complexes. A cheap and easy alternative to diatomaceous earth is to use small torn pieces of very coarse filter paper as a marsh.

Tab. 2.4: Sintered-glass (and porcelain) filter crucible sizes.

Porosity grade number	Glass porosity in μm (pore size index)	ISO designation
0 (very coarse, fast filtering) (±190 μm)	160–250	P250
1 (coarse) (±60 μm)	100–160	P160
2 (medium) (±30 μm)	40–100	P100
3 (medium) (±15 μm)	16–40	P40
4 (fine, slow filtering) (±2 μm)	10–16	P16

Cellulose filter papers are used in general laboratory filtrations, where it is only required to remove the solid matter without any further treatment of the solid matter or residue. The filtrate is then further analysed. If further treatment or analysis of the solid matter is required, then an appropriate filter paper must be used, such as ashless acid-resistant type.

Sintered-glass crucibles (also known as fritted glass filters, filter crucibles or Gooch crucibles) are used where the solid matter has to be removed from a liquid and also dried or burned (maximum temperature about 400 °C depending upon quality of the borosilicate glass) and weighed. Where higher burning (ignition) temperatures (>400 to 1,000 °C) are required then use porcelain glazed Gooch crucibles with a fibrous mat (or its equivalent).

Buchner funnels are used where the solid matter has to be removed and further processed, or for difficult to filter suspensions which require vacuum or pressure to assist in the filtration or separation.

How to select the correct filter paper?

There are many makes and types of filter papers (such as cellulose, cellulose nitrate or cellulose acetate) available, each with its own specific use. Always check the printing on all sides of the box of filters to note any manufacturer's recommendations, or other critical information.

The selection of the right filter type, paper, cellulose acetate, glass microfibre, phase separation paper or membrane filters, critical if accurate and precise analytical work is to be undertaken.

Filtration separation can be influenced, if the incorrect type is used, due to:
- high ash content of paper (e.g. contamination of ignited residues, hence, use ashless papers);
- degradation of the paper whilst still in use (filtering acidic or caustic suspensions using general-purpose papers), i.e. the degree of wet strength of the paper; hence use hardened paper;
- degradation of liquid (sample), or solvent evaporation, due to slow speed of filtration;
- whether it is necessary to use pleated paper, i.e. fluted or not;
- chromatographic separation (on the paper) of some components of the sample liquid;
- too large a pore size results in loss of particles, precipitates or sediment;
- qualitative analytical work where the analysis only requires whether the analyte is present or absent;
- quantitative analytical work where the analysis requires the accurate determination of the amount of the substance sought. Hence, use a hardened ashless-type filter paper for filtering out precipitates for gravimetric analysis.

It is recommended to always use hardened ashless cellulose papers when doing quantitative analysis. For qualitative analysis, any low-cost paper would be suitable.

The diameter of the filter paper selected and hence the size of the filter funnel used will depend upon the volume lots of liquid to be filtered.

Using pleated (fluted) or unpleated paper is a matter of personal preference of the analyst. However, the paper should be flat folded for greater gravity pull, without any spaces are air pockets between the funnel surface and the paper. To place a filter paper correctly in a funnel, tear a corner, then fold flap over; as shown in Fig. 2.5 below.

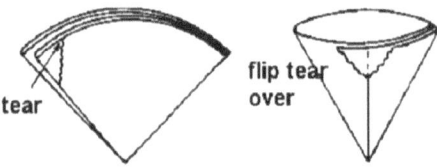

Fig. 2.5: Filter paper fold.

To separate two liquid phases, the analyst must select special papers such as Whatman 1PS or Sartorius 480. These papers are coated with silicon to render the paper hydrophobic, thus it retains the aqueous (water) phase and allows the organic phase (usually the solvent) to pass through, for further processing of the filtrate. This type of separation (filtration) is useful when the analyst prefers not to use glass separating funnels, because of the hazard of pressure build up in the separating funnel.

Pore sizes
This is an important consideration as it affects the size of the particles being retained as well as the speed of the filtration.
– Nanofiltration is from 0.001 μm particle size to 0.01 μm particle size;
– Ultrafiltration is from 0.01 μm (1 nm) size to 0.1 μm (100 nm; size viruses; about 1,000 molecular weight);
– Microfiltration is from 0.1 μm (100 nm) size up to about 10 μm (10,000 nm; size bacteria);
– Macrofiltration is from 10 μm (10 μm) to 100 μm (100 μm; particle filtration).

It is usually up to the analytical chemist's experience and knowledge that determines the procedure and type filtration media used for the type of analysis to be undertaken.

If in doubt, consult the manufacturers' catalogues on filtration products and their properties.

How to select the correct size filter funnel?
Select the paper of diameter that is less than twice the diameter of the funnel (a selection guide is shown in Tab. 2.5 below). The paper should not protrude beyond the rim of the funnel; otherwise, some of the liquid could be lost by the overflow of the liquid being filtered. If the paper is too large, then just trim off the excess paper with scissors.

Tab. 2.5: Filter paper and funnel sizes.

Diameter of paper, in mm	Maximum capacity of filter, in mL	Filter funnel size, in mm of top diameter
50	5	35
70	8	45
90	10	55
110	20	70
125	30	80
150	70	90
185	130	110
240	300	170
320	400	180

Procedure for filtering a suspension or sediment in a liquid (Fig. 2.6)

These steps below are necessary to promote the **filtration by gravity** of the liquid:

- Fold the selected filter paper (in either quadrants or fluted shape) as shown in Fig. 2.2 and place the folded paper in the correct size funnel.
- Wet paper thoroughly with distilled water (if an aqueous liquid is to be filtered, otherwise wet with the solvent used); discard these initial washings if the filtrate is to be retained for further testing.
- Press the wet paper to the walls of the funnel with a clean glass rod (aka policeman), so that there are no air spaces between the paper and the walls of the funnel. This is necessary as an aid to quicker filtration by the added suction or pull.
- Pour first the supernatant phase of the liquid slowly into the funnel without allowing the liquid to run down the outside of the beaker, by using a glass rod (policeman) to allow the liquid to run down the rod and directly into the funnel. If the liquid still tends to run on outside of beaker below the lip, then apply a small smear of silicone grease to the underneath of the lip/ridge of the beaker.
- Try to keep the stem of the funnel filled (as this assists the gravity pull of the filtrate drippings), but do not overflow the funnel.

If the test method requires the **residue** in the liquid to be tested, then transfer quantitatively any residue in the beaker by using a policeman (glass rod) and gently flushing the residue into the filter paper using distilled water (or the solvent, whichever one is applicable to the test). Wash paper and its contents thoroughly at least six times with water or the solvent. Discard filtrate and washings.

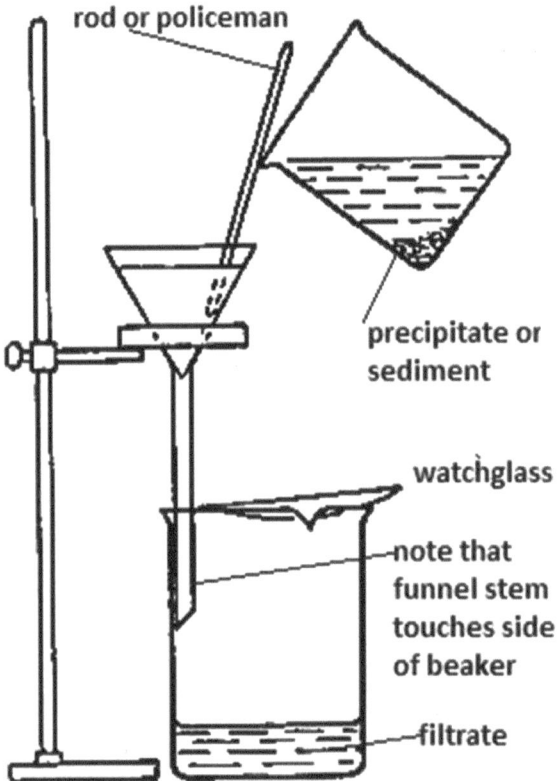

Fig. 2.6: Filtration.

If the test method requires the **filtrate** to be tested, then proceed carefully as stated in the test method without washing the filter paper, unless otherwise stated, in a standard test method. Select the correct type of paper for its intended use, by referring to Tab. 2.6 below:

Tab. 2.6: Whatman filter paper properties.

Whatman number	Filtration rate	Particle retention size (µm)	Some general uses
1	Fast flow	11	Qualitative; air pollution
2 or 2 V	Medium	8	Qualitative; coarse
3	Medium	6	For Buchner funnels, vacuum/pressure
4	Fast flow	20–25	Qualitative; air pollution

Tab. 2.6 (continued)

Whatman number	Filtration rate	Particle retention size (µm)	Some general uses
5 or 6	Slow	3	Fine particle filtration, vacuum/ pressure
40	Medium flow	8	Gravimetric analyses, AAS solutions, air pollution
41	Fast flow	20–25	Large particles and gelatinous precipitates
42	Slow flow	2.5	Small particles such as barium sulphate ppt.
43	Medium	16	Soils and foodstuffs
44	Slow	3	Fine particle filtration
50	Slow	2.7	Suitable for dilute acid/alkalis
52	Medium	7	For Buchner funnels, vacuum/ pressure
54	Fast flow	20–25	Not suitable for acid/alkalis
91 (Student Grade)	Fast flow	>10	E.g. Sugar tests or schoolwork
113 V (crepe paper)	Fast flow	30	Qualitative; wet-strengthened
540 (acid-resistant hardened paper)	Medium flow	8	Same as #40, metals, acid solutions, most common use
541 (acid-resistant hardened paper)	Fast flow	20–25	Acid/alkali solutions, proteins and fibrous foodstuffs
542 (acid-resistant hardened paper)	Slow flow	2.7	Same as #42, acid solutions
1 PS (silicone coated)	–	Water repellent	Aqueous/organic phase separations
GF/A (glass fibre)	–	1.6	Air pollution
GF/B (glass fibre)	–	1.0	Very fine particles
GF/C (glass fibre)	–	1.2	Water analyses
GF/D (glass fibre)	–	2.7	Coarse particles
GF/F (glass fibre)	–	0.7	Very fine particles

where V indicates pre-pleated or folded papers.

It is recommended to always use hardened acid-resistant ashless filter papers when doing quantitative analytical work, e.g. Whatman #541 or its equivalent.

Whatman Filter Papers and their approximate equivalents are given in Tab. 2.7.

Tab. 2.7: Filter paper equivalents.

Whatman number	MN (Macherey–Nagel)	Advantec MFS	Whatman S&S (Schleicher & Schuell)
1	645 or 615	2 or 231	591-A or 597
2	616 md	232	593
3	618	131	818
4	617	1	604
6	–	3 or 131	593-A
40	640 m	5B	589 white
41	640 we	5A	589 black
42	640 de	5C	589 red
43	640 m	7	589 green
52	1,672	–	1,574
54	1,670	–	1,573
114	1,670	–	410
540	1,640 md	–	1,506
541	1,640 we	–	1,505
542	1,640 de	–	1,507

POSTLIP (HRH) manufacture a student grade filter paper for ordinary work.

Chapter 3
Planning the analysis

Keywords: plan, selection test procedures

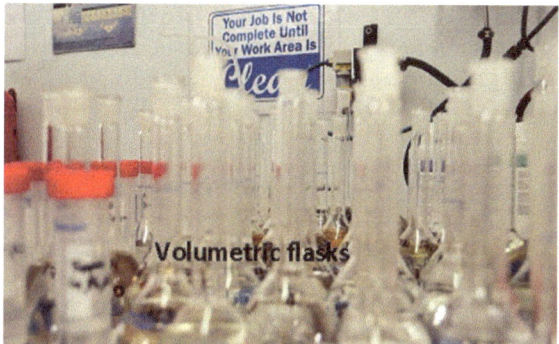

Fig. 3.1: Lab glassware preparation.

This chapter outlines most of the basics that have to be considered when undertaking analytical procedures or measurements.

Before starting any analysis, the analyst needs to know the following:

- What is the goal or **purpose** of the analysis?
- Was there a **sampling plan**?
- **How** was the sample **taken** and from what (statistical) population?
- Will the obtained analysis data be statistically **representative** of that population?
- Does the sample require freezing or refrigeration or any other type of **preservation**, before the analysis is undertaken?
- **How** was the sample transported or stored; or was there proper custodianship, tracking or chain of command, in place for the sample to prevent any chemical or physical changes, contamination or tampering, of the sample from the time it was taken to the time the analysis is started? Basically, the **integrity** of the sample.
- Is the sample container or package correctly and legibly labelled and **sealed**?
- **How many** samples are expected and what is the **frequency** of sampling? Refer to Fig. 3.1 as example of how many flasks to prepare.
- What procedures are in place if the integrity of the sample is **suspect**?
- What type of **sample matrix** or item is to be analysed?
- Is there any special sample **pre-treatment** required?
- Are there any possible chemical impurities present that could **influence** the selection of the test methodology?
- What is the degree of precision and accuracy **required**?
- **What** research, analyses or testing needs to be undertaken?

https://doi.org/10.1515/9783110721201-004

- **How** is the **analysis** to be done?
- **Where** is the analysis or testing to be done; on factory site, in the field, in the laboratory and so on?
- What **range** of quantities/concentrations can be expected for the analyte?
- **When** is the analysis to be done? Is there a time limit?
- Are all the necessary **resources** available at the laboratory to undertake the analysis or experiment?
- What **format** is required for the laboratory report; research thesis, certificate of analysis, certificate of compliance, forensic report as an expert witness and so on?
- What, if applicable, are reference standards or **specifications** available?
- Are there quality control (**QC controls**) or certified reference materials available, similar to the sample or item?

The answer to these questions will decide which approach or what techniques the analytical chemist must use in order to obtain the required information or computer data. This is known as the **sampling plan**.

Problems that can arise with **storage** of field samples include:
- Loss due to adsorption of the analyte onto the walls of the sample container
- Contamination due to desorption or leaching of potential interferents from the storage container
- Matrix changes due to microbial degradation for liquid aqueous samples, or oxidation/reduction and light/heat decomposition
- Water samples should be preserved with nitric acid to a pH of about 2; volume sample and volume preserved sample must be recorded, for test result correction calculations

Another important consideration is the type of measuring instrument or laboratory apparatus to be used will depend upon the expected level of the analyte; volumetric analysis for high concentrations (% levels) or spectrophotometric analysis for low concentrations (ppm levels).

Approximate **detection limits** of some instrumentation:
- From 1 ppt to 0.1% (1,000 ppm) for inductively coupled plasma (ICP) spectrometers
- From 0.1 ppb to 0.1% (1,000 ppm) for atomic absorption spectrometry (AAS) instrument
- From 0.1 ppq to 1% (1,000 ppm) for gas chromatography (GC) and high-performance liquid chromatography (HPLC) instruments
- Lower ranges can be obtained by attaching in line various extra detectors, such as GC-MS (mass spectrometry)

The analytical chemist must be aware of the **linear range** (working range or method linearity) measurement capabilities of the instrumentation that is being used, as well as its detection limit and limit of quantification.

A recommendation on how to **analyse an unknown sample** is to investigate further the following:
- What is its history?
- Is it radioactive?
- Are there any inherent toxic properties?
- Is it subject to self-ignition (spontaneous combustion)?
- Is it subject to explosions on heating or shocks?
- Is it:
  - animal (yeast, mould and biomass);
  - vegetable (organic matter);
  - mineral (inorganic matter)?

Based on answers to the above questions, the analytical chemist must decide what scheme of analysis to use for the unknown sample matrix, upon what degree of accuracy is required; then follow through with a literature search (Google search) for any refereed published test methods or sample preparation procedures which could be used, given the laboratory facilities on hand.

For **halides in organic substances**, use combustion methods; for **halides in inorganic substances**, use emission spectrographic, X-ray or EDX (aka energy-dispersive X-ray analysis (EDAX)) methods. If wet chemistry is to be used, then fusion procedures with selected fluxes must be used to obtain the unknown substance into a soluble form.

A general **guideline** to **the selection of analytical testing procedures**:
- For **percentage levels** of cation and anion analysis, use wet chemistry, such as gravimetric or volumetric analysis
- For **traces** of cations and anions use spectrophotometry, such as AAS, ICP and UV–vis spectrophotometry
- For **organic analysis** use chromatography such as thin-layer chromatography, GC, HPLC and ion chromatography
- For calcium, magnesium, chlorides and fluorides, use **ion-selective electrodes**
- For **unknown hard or refractory materials** such as rocks and ores, use X-ray techniques such as EDAX or X-ray fluorescence to identify the elemental composition of the material
- For **species identification** of the different oxidation states of anions and cations, use electrochemical techniques, such as voltammetry (current–voltage relationship during electrolysis) and amperometry
- For **functional-group** analysis, use infrared (IR) absorption, near-IR absorption, UV–vis absorption, nuclear magnetic resonance, mass spectrophotometry, polarography, Raman spectroscopy and general wet-chemical methods
- Selection on **speed and ease** of use or convenience of the method or testing process selected

– Availability of **resources**
– Amount of **sample available**; use of destructive methods or non-destructive testing procedures

The ultimate goal of every analytical chemist is to be able to report the most conclusive results to an analytical problem or routine samples, in the shortest length of time.

Chapter 4
Sampling

Keywords: sample plan, harvesting the sample, sampling equipment, sample preservation, primary sample, composite sample, analytical sample, decomposition, acids, wet digestion, fluxes, fusions, sample masses

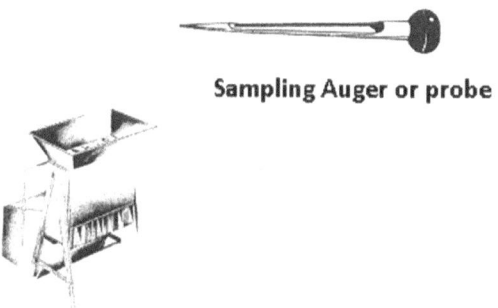

Sampling Auger or probe

Riffler or splitter

Fig. 4.1: Sampling.

How to sample or harvest items and how to prepare them for analysis is discussed in this chapter.

What is a *sample*?
A **sample** (aka an item) is a small separate part (e.g. a spoonful of sugar) of something exhibiting the same physical and chemical properties (e.g. of sugar) of a larger part known as the population (e.g. a 50 kg bag of sugar).

The sampling plan
The analytical chemist must devise an appropriate **sampling plan** based on appropriate statistical methods.

Sampling is also referred to as **harvesting**.

There are basically three ways to sample a consignment of goods or a product:
- the **easy** way (without any preplanning of how to sample), which leads to unreliable, non-reproducible sample and inaccuracy of final laboratory test results;
- the **quickest** way (taking short cuts and not sampling correctly from the sampling point), which leads to unreliable and inaccuracy of final laboratory test results;
- the **statistically correct** way, which leads to reproducibility, repeatability, accuracy and credibility of the final laboratory report, but it may not be the easiest or quickest procedure.

https://doi.org/10.1515/9783110721201-005

Many published sampling plans and procedures for various types of populations and sample matrices are available on the Internet such as ASTM, AOCS, BS, JIS, GATT, ISO and also in company standard operating procedures.

The custodianship or stewardship of the sample (i.e. the test item) is critical to ensure that the laboratory test results are not compromised in any way whatsoever before the analysis is undertaken. The integrity of the sample or item to be tested is critical.

Thus, if the laboratory analyst is also required to take a sample, then the analyst must follow the correct procedures because a **test result can *never* be more precise or accurate than the sample itself**, irrespective of how precise, automatic or expensive the laboratory measuring instruments used is.

Representative samples are taken by a systematic documented procedure and must be obtained by using such techniques that take into consideration the following factors:
– the type of product, consignment, substance, solid, gas or liquid;
– the particle size distribution if a product is in a solid, crystalline or amorphous powder form;
– the composition of product, that is, is it heterogeneous or homogeneous?
– the manner in which it is stored, e.g. storage tanks, mixing tanks, steel drums, plastic barrels, jumbo bags (aka flexible intermediate bulk container bags), glass jars, vials;
– the chemical or other hazards, if applicable; consult its material safety data sheet; now known as safety data sheet;
– what characteristics or preliminary tests have to be undertaken by the laboratory; e.g. moisture, size grading or density?
– what preservation precautions have to be taken before testing is started in the laboratory?

The most important and critical factor is that utmost care must be taken when sampling to prevent any changes in the sample's composition, quality, chemical and physical properties and to prevent any contamination of the sample from the sampling tools used. In other words, the integrity of the sample must be considered throughout the process.

Sampling equipment
Descriptions and diagrams of the most common utensils used in taking samples (such as augers and probes) can be found in any laboratory equipment catalogue or online (Google).

Sampling tools or equipment must meet these basic requirements:
– They must be robust enough to withstand handling operations
– Easy to disassemble to clean all parts of the equipment

- All parts of the equipment must be made of materials resistant to the type of goods or products being sampled, such as fruit's inherent acidic properties or chemicals
- The tools or equipment must also conform to safety requirements and any other specified specifications pertaining to the subsequent analysis
- For air or gas sampling these must be taken with special aluminium bags that contain septum plugs (especially useful for gas chromatography (GC) and high-performance liquid chromatography (HPLC) testing)
- For forensic specimens, use unreactive material for the sample containers, such as glass jars, aluminium cans, or polytetrafluoroethylene vials.

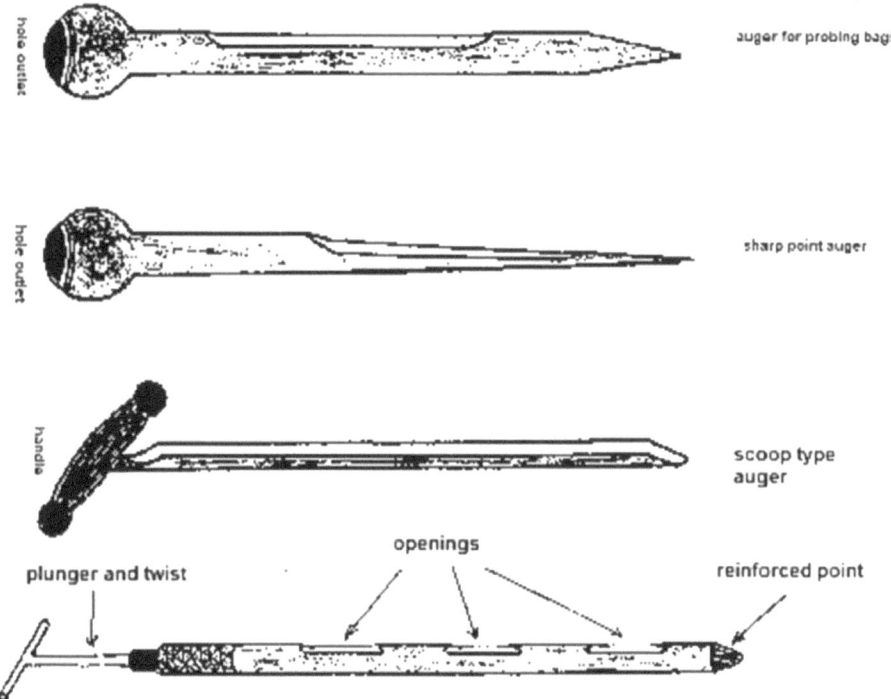

Sample preservation procedures

These must be considered, if any, before any transportation, storage or analysis is undertaken.

- Water sample may be preserved by the addition of **acid** or **preservatives**;
- **Photosensitive** samples must be taken quickly and stored in amber bottles or be placed inside a black bag or box;

- Samples sensitive to **air spoilage** or oxidation must be packed or filled in such a way that there are no air spaces inside the sample bottle or pack.

Other factors to consider are:
- What type of sample is it?
- What type of sample container is used, e.g. glass, plastic or metal?
- What type of population has the sample been harvested from?
- Does the primary sample need to be stored before submitting to the laboratory?
- How long will it take to transport the sample to the laboratory?
- How soon after receipt of the sample will the laboratory be able to start testing?
- Is there any custodianship standing procedures (aka **chain of custody**) to be followed, such as tamper proof sealing of the sample?
- Does the sample require to be kept frozen or chilled before testing?
- Does the sample need to be kept vacuum sealed or under an atmosphere of nitrogen to prevent degradation?
- Is there a tracking or sample recording/logging mechanism in place for the sample before it reaches the laboratory?
- Are there records or logbooks of what, if any, preservation procedures that were undertaken?
- Is it necessary to complete a sample submission form including notice of any preservatives added?

What is a **primary sample**?
This is generally the first representative sample, or gross sample, that is taken from a lot and it could consist of a number of smaller samples taken from the same lot to be composited (i.e. added together) to form the primary sample (a composite sample).

There are various ways of how to collect or take a sample:
- Grab sampling, where a sample is taken quickly and frequently at a certain place, such as a conveyor belt.
- Random sampling, where no pre-calculated proportion of the population is taken and at no particular part, area or place.
- Automatic or conveyor belt sampling; similar to the grab type but is done by robotic or mechanical means.
- Composite sampling, where several samples (sub lots) are taken at different places then mixed together on site, or alternatively taken to the laboratory where the composite is prepared.
- Sampling of metals can be done by sawing, drilling or filing; with precautions not to contaminate the sample with the metals from the tools or equipment used. Analytical sample can then be further prepared by using ball mills or hammer mills to grind or pulverise the test item (sample).

The primary sample is then reduced in volume, mass or size, according to certain procedures, such as for solids using a **riffler** or by the method of **cone and quartering** of particles.

Cone and quartering is basically mixing the sample particles into a heap shaped like an inverted cone and then flattening out the cone into a pie (pizza) shape, followed by dividing the pie into quadrants, discarding diagonally opposite quadrants and then remixing the remaining two quadrants; repeating this exercise until suitable volume or amount of sample is retained for the laboratory tests.

A general guide is:

- for homogeneous substances such as orange juice, cement and sand:
  take about 250 g sub-lots;
- for non-homogeneous (heterogeneous) substances such as feeds, mixtures and ores:
  take about 500 g sub-lots.

Finally, a **composite** is made of all these sub-samples (sub lots) to form the primary sample. This primary sample can then be reduced in volume or size (to make a secondary sample) by using:

- if pellets, grains or powder, then riffler (Fig. 4.1), or by cone and quartering; or
- if a liquid, then by simply taking an aliquot (sub-sample) out of the well-mixed liquid primary sample.

This **secondary** sample is then reduced in size or amount from the primary sample and may need further processing, e.g. milling of particles, to obtain the suitably pulverised final analytical sample.

The **analytical or test sample** is then used for all the required tests to be undertaken. Note that this sample must be stored in a clean air-tight container (because the sample might either absorb moisture (or oxidise) with time, or it can also dry out with time). Proper preservation techniques must be used depending on the sample's chemical make-up or matrix.

The nature of the sample will determine what type of container to be used, e.g. for orange juice (i.e. an acidic type sample); do not use an iron or tin container, but use an aluminium container.

Preparation of the analytical test sample

What is the **analytical sample**?

This is the **final homogeneous and representative sample** that has been prepared from the primary and any secondary samples (if necessary, it has been milled to a powder form) to create the amount of sample that is given to the analytical chemist for testing.

Sample preparation must be treated in such a way that a homogeneous representative sample can be produced via specific preparation techniques, such as addition of chemicals or catalysts, combustion and dry or wet ashing, filtrations, centrifuging, acidic digestions, solvent extractions, distillations, ion exchange or chromatographic separations, fusions with a specific flux, ultrasonic baths or microwave radiation.

This is another most critical part in any analytical process or experiment.

Consideration must be taken into account for an applicable procedure to isolate or separate the analyte from any interfering substances that might be present in the sample makeup.

Methods of **separation prior to analysis**:
- Precipitation (gravimetric)
- Fractional distillation
- Selective solvent extraction
- Ion exchange
- Adsorption chromatography
- Partition chromatography
- Electrodeposition
- Electrophoresis
- Osmotic separation (membrane filtration)

All separation techniques involve one or more chemical equilibria; hence, the degree of separation achieved can vary greatly depending upon the experimental and partitioning (extraction) conditions.

For **liquid samples**, immediately before testing, shake the bottle or container carefully but thoroughly to ensure the re-dissolution of any sediment that might have formed during transport and storage – be aware of any possible pressure build-up! *Do not* shake the bottle or container if the sample has to be tested on a "settled solids" or "suspended solids" basis.

For **viscous samples**, such as molasses, fats, heavy fuels, lube oils and waxes, loosen the lid of the container and warm in the incubator or oven at a low temperature just until mixture can be safely mixed or stirred to form a homogeneous blend – *do not overheat* as this could irreversibly change the chemical and physical properties of the sample! And could also start a combustion process!

For **solid samples, such as coal, animal feed pellets, sugar and grains,** if more than sufficient amount of sample has been received by the laboratory (e.g. from the preparation of the sub-lots), then reduce the amount by first breaking any lumps or pieces of sample larger than about 20 mm into smaller pieces, until *all* the sample received is of uniform particle size of about 10 mm or less – *do not do this* if size grading (or particle size distribution) of the sample is required. If sufficient sample is available, then retain a representative portion of it for its size grading and prepare the rest of the sample for analysis.

Once all the large lumps have been milled or crushed to below 10 mm particle size, reduce the amount of sample, if necessary, by using a sample riffler of by using the technique of cone and quartering as described above to a quantity sufficient for the analytical tests.

During sample preparation, DO NOT contaminate the sample in any way! For example, if the sample is to be tested for its iron content, then do not use a metal hammer or metal apparatus to break any lumps. Use a robust non-iron apparatus. If a metal apparatus is used, be aware of sparks.

The sample is then reduced by milling and so on, if necessary, into a powder form of about 1 mm in particle size for the final analytical tests.

Note that if a moisture test is to be done on the sample, then the above procedures MUST be done as quickly as possible to avoid changing the moisture level in the sample by heat caused by the friction of the grinding or milling process.

If the sample is a liquid and the test substance (analyte) is in an unsuitable chemical matrix, then one or more of the following steps must be undertaken:
– Oxidation by reacting sample with an oxidising agent, e.g. reaction with bromine water;
– Reduction by reacting sample with a reducing agent, e.g. reaction with zinc;
– Chelation (formation of a metal complex), e.g. addition of EDTA;
– Decomposition, e.g. acid digestion and ashing.

The analyte elements, such as nitrogen compounds or oxides, halogens, sulphur compounds and phosphorus compounds, in an organic sample or substance are usually determined by combustion techniques. This can be done by instrumentation, such as the CNH analyser, or by dry digestion or wet ashing of the sample under an oxidising atmosphere.

Decomposition of the sample can be undertaken by using various acids, caustic, water or organic solvents. Most analytical chemical techniques require the sample to be in a liquid form prior to the analysis.

Examples are:
– Water extracts by heating, boiling, refluxing; water is the most common solvent
– Acid extraction:
    – Hydrochloric acid (HCl) is a common useful strong acid (aka spirits of salt), but very corrosive and volatile (fuming); it is also a mild reducing agent.
    – Nitric acid (HNO_3) is an oxidising strong acid for digesting organic matter; concentrated fuming nitric acid is very hazardous and should be used with extreme care. Not suitable for aluminium and chromium due to their passivity properties.
    – Sulphuric acid (H_2SO_4) is a strong acid (also, it boils at high temperatures, about 340 °C) that attacks organic matter, dehydrates it and then oxidises it into carbon. This acid reacts violently with water! Sulphuric acid of higher than 100% H_2SO_4 is known as oleum ($H_2S_2O_7$) and is a very hazardous fuming acid that is used in manufacture of various chemicals.

- Aqua regia (royal water) is so called because it dissolves the noble metals like gold; it is a mixture of concentrated hydrochloric acid and nitric acids (in ratio 3:1). Useful for dissolution of some refractory materials.
- Hydrofluoric acid (HF) for silicate-type minerals, porcelain, glass, ceramics and similar materials. It causes severe invisible burns through the skin and attacks the bones.
- Perchloric acid (HClO$_4$) is a very dangerous corrosive and explosive chemical, especially in contact with organic matter, but useful in controlled acid digestions. **Do not** allow a boiling solution with perchloric acid to boil down to dryness, otherwise an explosion will occur! It attacks iron alloys and stainless steel samples.
- A mixture of perchloric and nitric acid is less dangerous than perchloric acid on its own!
- Addition of hydrogen peroxide to mineral acids, such as hydrochloric acid, hastens the decomposition of organic matter in samples.
- Ashing, wet ashing with acids; followed by ignition in a muffle furnace.
- Dry ashing using a Bunsen burner or muffle furnace.
- Alkali extraction, such as sodium hydroxide, for the dissolution of amphoteric materials such as aluminium alloys.
- Solvent extraction for organic type materials and polymers.
- Fusion using various fluxes, where the flux chemical is generally added in amount of 10 times that of the sample weight. Used widely in ore assays.

Decomposition of the sample can also be undertaken by using various other processing steps, or in combination with the above chemicals, such as:
- Ultrasonic bath with temperature control
- Microwave digestion
- Oxygen bomb digestion vessels, such as the Parr unit
- Fire assays for noble and refractory materials
- Stomachers (paddle blenders) for soft-like or pastry-like materials, such as foodstuffs and biological specimens
- Rotating ball mills for grinding or pulverising minerals, ceramics and similar materials
- Hammer mills for shredding or crushing materials
- Grinders (for hard substances) and blenders (for soft substances)

A general procedure for difficult digestion of solids, such as, foodstuffs, pet foods and fertilisers, where the analytes required are metals (minerals), is a **wet digestion** with **sulphuric acid** and **hydrogen peroxide** (sulphuric acid is used because of its **high boiling point** compared to the other acids), as below (always wear lab safety goggles!):

1. If necessary, pulverise sample to about 2 mm particle size; dry the sample first if it contains more than 5% moisture content
2. Mass out (weigh out) 1–3 g of the dry sample into a reflux flask (with condenser) or Kjeldahl flask
3. Add slowly and carefully 3–5 mL concentrated sulphuric acid by allowing any reaction to subsidise
4. Heat gently on a hotplate or heating mantle (about 200 °C mid-setting) and then add slowly and carefully a few drops of hydrogen peroxide until the solution is more or less clear
5. Heat further to about 500 °C (max setting or high) and add about 2.5 mL peroxide slowly, allow all reactions to subside; maximum time period should be about 45 min; do not heat to dryness
6. Cool, remove condenser, add carefully few millilitres water and transfer all contents to a 100 mL clean volumetric flask, make volume to mark with water. The (clear sediment-free) solution is now ready for metal analysis
7. Always run a reagent blank simultaneously with the samples by adding same quantities of acid and peroxide as that added to the samples and following all the same procedures. Any metal detected in the reagent blank is then subtracted from the sample readings.

The following acids can be used for some other difficult type samples:
– Aqua regia acid (3 HCl:1 HNO_3) for sulphides
– Perchloric acid ($HCLO_4$) for steels and similar metals
– Perchloric acid + phosphoric acid (H_3PO_4) for chromites

If any sample acid digestion procedures, especially for minerals and ores, does not produce a sediment-free liquid for metal testing, then a **fusion or fire assay** has to be undertaken.

A **fusion process** is where another material (**flux**) is added to a sample in a specified crucible, mixed well, then heated to high temperatures (950 °C or higher for refractory materials), until molten state is formed. The hot liquid melt is gently swirled slowly for a few minutes to give the sample a chance to dissolve completely in the liquid flux, until a uniform clear molten liquid, free of insoluble particles, is formed. This is allowed to cool slowly; the appearance of the melt should be that of glass.

This process should not take more than 15 min. At this stage, the melt should be allowed to mature for several hours or overnight.

This process converts the insoluble metallic analytes into a complex inorganic compound that is soluble in acids and hence can be used for further measurements, vis X-ray fluorescence (XRF), atomic absorption spectroscopy (AAS), inductively coupled plasma (ICP)-OES (optical emission spectroscopy) or ICP-MS (mass spectroscopy).

Basically, a **flux is an additive** which permits the basic components of a glaze to fuse together at a **lower temperature** to form a **homogenous** mass.

Disadvantages of decomposition of a sample by fusion are:

- Contamination of the sample by the large amount of flux added (can be avoided by running a reagent blank)
- The final aqueous solution will have a high salt content which could cause chemical and spectral interference upon further analysis of the melt
- The high temperatures required for the fusion or melting could cause loss of sample volatiles
- The crucible is inevitably attacked by the flux, leading to cracks, permanent staining/contamination of the crucible

Note: The type of flux to be used for a fusion is determined by the nature of the flux and sample matrix.

Generally, use the **acidity and alkalinity equilibrium principle**, namely a basic flux, to react with an acid-type sample or an acidic flux with an alkaline-type sample (Tab. 4.1 and 4.2).

Tab. 4.1: Fluxes.

Flux	Crucible	Flux: sample ratio	Fusing temperature (°C)	Type sample
Basic fluxes: Sodium carbonate, lithium borates	Nickel, platinum	10:1	700–1,000	Acidic: Iron oxides, phosphates, sulphates, silicates, glass, porcelain, clays
Basic fluxes: Sodium hydroxide, potassium hydroxide	Nickel, platinum	5:1	400–600	Acidic: Silicates, minerals
Basic fluxes: Sodium hydroxide or potassium hydroxide	Nickel	10:1	330	Silicates
Acid flux: Boric oxide, acid fluorides,	Platinum	5:1	650	Metallic oxides
Acid flux: potassium pyrosulphate	Nickel, platinum, porcelain	20:1	450–900	Basic: iron oxides, basic metal oxides, rocks, non-metals, titanium oxide, porcelain, insoluble oxides of aluminium

Tab. 4.1 (continued)

Flux	Crucible	Flux: sample ratio	Fusing temperature (°C)	Type sample
Acid or base flux	Nickel or platinum	10:1	800–900	Amphoteric: aluminium oxides, tin oxide, chromium III oxide, iron III oxides
Oxidising flux: sodium peroxide	Iron or platinum	10:1	600	Oxidisable substances: chromite
Oxidising flux: sodium peroxide + sodium carbonate	*Do not use platinum*			Samples requiring a strong oxidising agent, metals, silicates, ceramics, waxes
Basic flux: sodium perborate + sodium carbonate	Platinum	10:1	950–1,100	Refractory minerals: chromite, titanium dioxide, aluminium oxide, ferric oxide
Basic flux: sodium tetraborate + sodium carbonate	*Do not use nickel crucibles*	8:1	900–1,000	Silicates

The maximum temperature to be used must be such that it will not attack the container (crucible), or decompose or break down the chemical composition of the flux itself.

Basically:
- **Platinum** crucibles are generally used for most fusions, **but never** for nitrogen-containing samples or fluxes, peroxides, chlorides or sulphites.

  Platinum ware can be **cleaned** by:
  - boiling in 10% hydrochloric acid (beware of the HCl fumes which are corrosive to the laboratory environment);
  - undertaking a blank fusion with sodium hydrogen sulphate;
  - scrubbing gently with some aggressive powder, such as sea sand, or a sodium metasilicate cleaning paste (avoid cementation).
- **Nickel** crucibles are used for sodium hydroxide fusions, but never for sulphur or nitrogen containing samples or fluxes; also, they are generally not suitable for sodium peroxide, sodium tetraborate + sodium carbonate fusions; nickel is also reduced in presence of sodium peroxide and by nitric acid.
- **Zirconium** crucibles can be cleaned by boiling in dilute hydrochloric acid, but avoid any presence of HF.
- **Iron** crucibles are never used with acidic fluxes or acidic samples.

- **Porcelain** (glazed) crucibles are used when platinum crucibles are not available, but are subject to staining and cracking.
- **Graphite** crucibles can be used in place of metal crucibles, but possible reduction of the analyte sought could take place due to the presence of the carbon. Graphite crucibles are disposable which eliminates the need for cleaning. Graphite will breakdown over lengthy periods of high heat exposure.

Further information about cleaning crucibles can be found in Appendix XIX.

There are many flux blends sold by various suppliers and as such the analyst must diligently select the flux that is most suitable for fusing with the sample matrix at the flux specific temperatures without damaging, etching or reacting with the crucible's composition.

For ICP and AAS preparations, a hot dry block is more efficient digestion than a microwave oven for higher sample weights (or volumes) such as 10 g or more. Acid digestion is normally 1–2% nitric acid for ICP work. The acid concentration must match the solvent matrices of the Standards used.

The total dissolved solids (TDS) for AAS work should not exceed 10% to prevent burner blockages due to carbon particle build up. For ICP, the maximum TDS is 25%.

Tab. 4.2: Sample masses.

Technique	Minimum amount sample required
AAS	100 mg
FES	200 mg
FTIR spectroscopy	10 µg
Raman spectroscopy	10 µg
UV-visible spectroscopy	10 mg
Gas chromatography	1 mg
Chemical reactions and stoichiometry	1 g
ICP-OES	10 mg

Instrumental analysis

Many instrumental analyses are undertaken based on different electrochemical methodologies. These measure the potential (volts) and/or current (amps) in an electrochemical cell containing the analyte. The methods can be categorised according to which parts of the cell or electrode are controlled and which are measured. In order to control an electrical signal, the instrument needs to be able to measure it.

The four main categories are:
- **Potentiometry**, where the difference in electrode potentials (pd) is measured (millivolts)
- **Coulometry**, where the transferred electrical charge (coulombs) is measured over a certain time period
- **Voltammetry**, where the cell's current is measured during an applied alternating voltage (AC)
- **Amperometry**, where the cell's current is measured over a certain time period

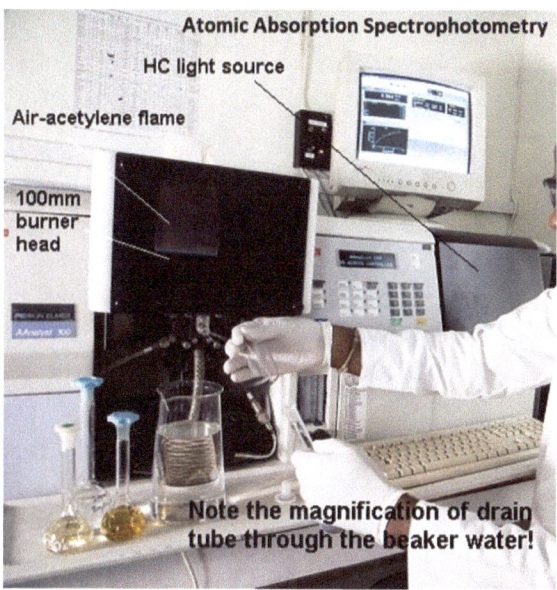

Fig. 4.3: Instrumental analysis.

Some complex and very expensive analytical instrumentation are in analytical chemistry, such as AAS, ICP, FTIR, GC-MS, NMR, IC, NIR, XRF, HPLC and many less common, but used in university research laboratories and government departments.

Many of the most common instrumental techniques are discussed in Chapters 5–14.

The result from an instrument is only as *accurate* as the correctness of that instrument's verification and calibration. Thus, no matter how *precise* and *reproducible* is the readout/display/spectrum/printout from the instrument, that value or data figures, can never be more accurate than the correct operational procedures by the laboratory, irrespective of how expensive or sophisticated the instrument and its software programme may be.

The **principal** here is the operation and correct use of digital (electronic) detection and measurement of levels of a substance in a complex media and controlled by instructions from computer programmes (software). Even simple pH measurements are now done by computerised instrumentation.

The problem which many analysts overlook is when it comes to compiling calibration graphs or curves, by using spreadsheets such as Excel or specific manufacturers' software. These computers programmed spreadsheets can give one a false sense of precision, when the analyst types in request for linear or trend lines, from a set of measured instrument standard calibration values.

Instrumental analysis

Chapter 5
Calorimetry

Keywords: heat energy, units, specific heat, heat capacity, cal, calorimeters, terminology

An outline of the energy principles and types of measurements that an analytical chemist has to undertake is given in this chapter. Equipment is discussed and some useful data tables are given. Terminology and units are included.

Calorimetry and thermogravimetric analysis measure the interaction of a substance with either absorbed heat or emitted heat.

The elements carbon, hydrogen, oxygen, nitrogen (Dumas), halogens, sulphur and phosphorus, in organic substances are best measured by the combustion process.

The **principle** is the measurement of the amount of heat energy transfers of mixtures, when together and exposed to temperature differences.

Reactions where substances **absorb heat** are known as **endothermic** and those that **release heat** are known as **exothermic**.

The **calorie** is a **unit of energy** and is **defined** as the amount of energy required to raise the temperature of 1 g of water by 1 °C.

The basic **principle** is where the measurement of heat energy lost by a material at high temperatures is equal to the heat energy gained by a material at low temperatures. This transfer of energy ceases when both materials are attained at equal temperature; that is at a state of **temperature equilibrium**.

Heat energy is measured in calories (cal or C), joules (J or kilojoules) or British thermal units (BTU).

1 cal = 4.1868 J
1 Cal = 1 kcal
1 kcal = 1,000 cal
1 BTU = 1,055 J
1 BTU = 252 cal

where kcal is kilocalorie; kJ is kilojoule; and Calories (with an upper-case letter) is regarded in the food industry as the same as kilocalories (Tab. 5.1 and 5.2).

The **energy system** must be isolated from any external heat or cold environment.

The system is dependent upon the **specific heat** factor, c, of each of the two bodies and their masses (m); this is known as the **heat capacity** (c × mass) of that body.

The equation is simply:

Mass of hot body × c of the hot body × temperature change = mass of cold body × c of the cold body × temperature change

https://doi.org/10.1515/9783110721201-006

Tab. 5.1: Energy values of major food types.

Carbohydrates, as monosaccharides = 3.8 kcal per 100 g = 16.5 kJ per 100 g
Glycitol = 3.8 kcal per 100 g = 16 kJ per 100 g
Proteins = 4.0 kcal per 100 g = 17 kJ per 100 g
Alcohol, as ethanol = 7.0 kcal per 100 g = 29 kJ per 100 g
Fats and oils = 9.0 kcal per 100 g = 37 kJ per 100 g
Sugar alcohols = 2.4 kcal per 100 g = 10 kJ per 100 g
Organic acids = 3.0 kcal per 100 g = 13 kJ per 100 g
Citric acid = 4.1 kcal per 100 g = 17 kJ per 100 g

where c is the specific heat value of that body.

Example:

The transfer of heat from a burning substance to that of the cooling water surrounding it.

Heat (energy) will always flow from a hot region to a colder region (never the reverse) until both are at the same temperature. This is denoted by:

$$Q = m \cdot c \cdot \Delta T, \text{ where } \Delta T \text{ is change in temperature.}$$

In the laboratory, the above principle of **thermal energy transfer** is used in **calorimeters** (e.g. in presence oxygen or a bomb calorimeter), which measures the calorific value (heat of combustion) of substances, such as foodstuffs, solid and liquid fuels, coal, coke and biofuel, when they are combusted under defined conditions.

The sample, of known mass, is placed in a pressure vessel which is enclosed in an outer jacket containing water (bomb calorimeter); the pressure vessel is under a rich atmosphere of oxygen and is electrically ignited by heated filament (thin fuse wire); the resulting explosion generates heat energy. This temperature rise is measured by accurate thermometers or resistance sensing devices, to an accuracy and precision of 0.001 °C.

The **calorific value** may be stated as the number of heat units liberated by a unit mass of a sample when combusted in a fixed volume of oxygen. The reaction is the oxidation of all carbon and hydrogen with oxygen to form carbon dioxide and water.

There are some factors that have to be taken into consideration which could cause the incomplete combustion (or side reactions) of the sample or the material of interest. Example is the heat liberated by the presence of any sulphur or chlorine compounds which may be present in the sample or material. A correction factor is undertaken by adding barium chloride to the water after combustion and determining the barium sulphate levels.

To **calibrate** or verify a calorimeter, a substance of known specific heat is used, such as benzoic acid with a heat of combustion of 6,318 cal/g. This is required to determine the energy equivalent of the calorimeter.

The calorific value can also be calculated mathematically from the individual calorie values of the components of the sample.

There are many types of calorimeters, both manual operation and automatic computer controlled:

- **Constant volume** calorimeter, such as the bomb calorimeter.
- **Globe electric** calorimeter, a simple yet effective piece of student apparatus. It consists of an inner container and the outer jacket container and has a 6 DC volt battery supply.
- **Adiabatic** calorimeters are where the heat of the reaction conditions is not held constant, but the temperatures are measured, without any heat entering or leaving the system; **bomb** calorimetry is used to determine the enthalpy of combustion.
- **Reaction** calorimeters are those that measures the heat evolved and the rate of heat evolution from a chemical process under defined conditions.
- **Constant pressure** calorimeter measures the change in enthalpy of a reaction occurring in a liquid solution at a fixed pressure.
- **Differential scanning** calorimeters are automatic instruments based on the thermoanalytical principle in which the difference in the amount of heat required to increase the temperature of a sample and a reference material is measured as a function of temperature.

Terminology

The molar heat of combustion is the heat released when one mole of a substance is completely burned (combusted).

Molar heat capacity is the heat capacity per mole of a pure substance. This can be defined as under constant volume or constant pressure.

Specific heat capacity (aka heat capacity) is the heat capacity per unit mass of a pure substance, denoted by c.

Enthalpy is defined by the equation:

$$H = E + PV$$

where H is the heat content of a substance, E is the internal energy of the substance and PV is the pressure and volume (usually taken as 1 atm pressure at 25 °C).

Gross heat of combustion (gross calorific value) is the quantity of energy released when a unit mass of combustible material is burned in a constant volume enclosure (e.g. bomb calorimeter).

Net heat of combustion (net calorific value) is the quantity of energy released when a unit mass of combustible material is burned under defined conditions and mathematical corrections are done to compensate for presence of moisture in the sample, carbon, sulphur and hydrogen containing compounds:

- Carbon bonds, all are combusted to CO_2
- Hydrogen bonds, all are combusted to H_2O
- Sulphur bonds, all are combusted to SO_2

Tab. 5.2: Specific heat.

Material	Specific heat capacity, c, at 25 °C, in J/g/c
Water, liquid	4.18
Alcohol	2.46
Ice @ 0 °C	2.01
Steam @ 100 °C	2.01
Air	1.02
Brass	0.38

Chapter 6
Chromatography

Keywords: stationary phase, mobile phase, column chromatography, gas chromatography, ion chromatography, paper chromatography, thin-layer chromatography, high-performance liquid chromatography, partition coefficient, polarities, liquid phases, van Deemter equation, detectors, Kovats retention index, McReynolds constants, evaluation methods, response factors

This chapter outlines many techniques and procedures on how to separate the parts of a whole substance, or item, or sample and their identification thereof, by various analytical procedures as stated below. The theory and various equations are not outlined below; but can be found in many reference textbooks and from the many instrument manufacturers' websites on the Internet.

All chromatography techniques depend upon the same basic principle, that is, the variation in the rates at which different components of a mixture will migrate through a stationary phase under the influence of a mobile phase.

When to use which technique depends on the matrix of the sample, the expected, if any, concentration levels sought, amount sample available and what kind of breakdown of the sample is required.

Supercritical fluid chromatography can be classed as a hybrid between gas chromatography (GC) and high-performance liquid chromatography (HPLC) employing a supercritical fluid as the mobile phase. These fluids are produced when gases and liquids are exposed to temperature and pressure changes which exceed their critical physical values or properties.

The field of chromatography is very wide and diverse. Many analytical chemists only specialise in one or more chromatographic fields, as consultants and are often called upon in litigation and forensic cases, to give expert evidence.

All forms of chromatography involve a **stationary phase** and a **mobile phase**. In gas–liquid chromatography (GLC, or **GC**), the mobile phase is a **gas** such as helium and the stationary phase is a high boiling point liquid adsorbed (coated) onto a solid. In most other forms of chromatography, the **mobile phase** is a **liquid** (e.g. thin-layer chromatography (TLC)).

Separation processes are used to decrease the complexity of the makeup of a substance or chemical compound.

The **principle** of the chromatography type of separation is the relative distribution (partition coefficient) between two immiscible phases. One phase is the **stationary phase** of a large area (such as a TLC plate or glass column) and the other is known as the **mobile phase,** such as a gas or fluid that passes slowly through the

https://doi.org/10.1515/9783110721201-007

stationary phase leaving behind in its wake adhesive-like parts (molecular groups or ions) of the substance.

In other words, chromatography is the collective term for a set of **laboratory techniques** for the **separation of mixtures and compounds**. The mixture or compound is dissolved in a fluid called the mobile phase (which may be a gas, a liquid or a supercritical fluid), which permeates its way through an immobile, immiscible material called the stationary phase.

The various constituents of the mixture or blend of compounds move through this material at **different speeds**, thus causing them to separate (because they have **differing solubilities** and mobilities in each specific mobile phase) into their individual components or species. This separation technique is based on **differential partitioning** (or solubility characteristics) between the mobile and stationary phases.

The **partition coefficient** is an indication of how much **concentration** is **distributed** between the mobile and stationary phases when a sample is injected into a chromatographic column.

The least retarded (fastest moving component) elutes first, that is, it appears first out of the column (or top of a TLC plate) and is detected by a specific type **detector** (e.g. thermal ionisation detector (FID) or thermal conductivity detector (TCD)) depending upon the nature of the eluate. The most strongly retained component elutes last from the column.

The reason why some components are retained for longer time periods than others is because of their difference in **polarities** (i.e. partition forces). Polarities of compounds vary from strongly polar to strongly non-polar compounds.

Retention behaviour is affected by the distribution of the solute between the **mobile phase** and that of the **stationary phase**.

In other words, the mobile phase (mobile phase is a carrier gas, such as nitrogen, hydrogen, argon or helium) moves through the stationary phase (stationary phase is normally types of solids such as silica) picking up the compounds (components) to be detected and measured.

The most critical component of an HPLC, ion-exchange chromatography (IEC) or ion chromatography (IC) or GC is its **column** which does the separation of the sample's components. The selection of the right column for the specific sample matrix and what components are sought is important in order to get adequate separation of the components (organic molecules or ionic species).

The various chromatographic techniques are classified according to their principles of separation.

- Liquid–liquid or **partition** chromatography involves a simple partitioning separation between two immiscible liquid phases, where one is a stationary phase and the other one regarded as the mobile or moving phase.
- Liquid–solid or **adsorption** chromatography involves a simple separation between two phases, where one is a solid stationary phase and the other one is the

mobile or liquid phase. The physical attractive surface forces are involved in the retention capability of the solid stationary phase.
– When ionic components of a sample are separated by selective exchange with opposite charged **ions** of the stationary phase; this is known as **IC**.
– **Gel permeation** chromatography (aka gel filtration or exclusion chromatography) is commonly used in biochemistry; it is similar to gel electrophoresis for the separation of proteins.
– GLC is one of the most common techniques used in chromatography. As it states, the mobile phase is a gas and the stationary phase is a solid. This technique is commonly referred to as **GC**.
– In **gas–solid** chromatography (**GC**), the separations are done by applying temperature gradients to the separation process.
– In HPLC, the separations are done by applying pressure gradients.

6.1 Column chromatography

Column chromatography is commonly used as a clean-up stage or purification stage of a sample (solution) for further analysis.

The principle here is the **differential adsorption** of the solutes (compounds) of a solution passing through a stationary phase, at different rates by gravity, which then separates the mixture into individual components. These components are then subsequently collected as eluates for further analysis or identification.

This process is based on the affinity of each component towards the mobile phase (generally a solvent, such as chloroform or methylene chloride) and the stationary phase (generally silica gel granules of uniform size 0.1–0.2 mm).

Note that the packed column should be kept moist, with the solvent before use, without any entrapped air bubbles present. Also, the column packing should be free of water moisture, hence, generally a plug of anhydrous sodium sulphate is used at the top of the column packing, as a pre-drying stage.

When slowly pouring the sample solution into the column, do not allow the column to run dry.

The eluates (or fractions) are then collected as and when stipulated (e.g. time intervals) in the analytical procedure.

6.2 Gas–liquid chromatography

A gas chromatograph consists basically of:
– a supply of a constant flow of carrier gas, normally nitrogen, hydrogen or helium, controlled by pressure gauges or valves and flow meters;

- a supply of pressurised detector gases (sometimes these are the same as the carrier gas), such as air and hydrogen for the FID detector;
- sample injector ports: direct on-column, flash vaporiser, split or splitless injection;
- sample injector splitter valves;
- detector ports;
- column oven with controllable temperature settings or temperature programming;
- columns, made from various materials, various lengths, various shapes and various internal diameters; capillary open tubular or wall-coated open tubular;
- suitable packed or capillary columns; packed columns can be of silanised glass or metal nickel or stainless steel);
- microlitre syringes and septums;
- computer and printer with chromatography programme software.

In most cases, the columns consist of semi-liquid chemicals (**stationary phases**) (Tab. 6.1 and 6.2) coated onto a solid chemical (**support phase**), through which a carrier gas passes (**mobile phase**) at a predetermined gas flow rate. The temperatures of the separations (partitions) in the column are dependent upon the maximum temperature that the stationary phase can withstand (before bleeding takes place in the column).

There are basically two types of columns used, namely **packed columns** and **capillary columns** (aka open tubular columns). These columns are coiled so that they can be of various lengths, but also fit inside a small oven box.

There are many mathematical equations where one can calculate and determine efficiencies such as **column selection** and **resolution**, carrier **gas flow rates**, partition coefficients and ratios, **van Deemter equation** (Fig. 6.1), **Kovats retention indices, McReynolds classification of stationary phases**, etc.

Tab. 6.1: Common GLC stationary (liquid) phases.

Trade name	Chemical composition	Maximum oven temperature (°C)	Polarity	Uses
Carbowax 20 M	Polyethylene glycol	250	Polar	Alcohols, esters, essential oils
OV-1	Polymethyl siloxane	350	Non-polar	Hydrocarbons, drugs steroids
OV-17				
Squalene	$C_{30}H_{22}$	150	Non-polar	Hydrocarbons
Bentone				

Tab. 6.1 (continued)

Trade name	Chemical composition	Maximum oven temperature (°C)	Polarity	Uses
DEG adipate		200	Semi-polar	
Triton X-100				
SP-2100				
Apiezon				

The van Deemter equation relates to the variance per unit length of a separation column to the linear mobile phase velocity by considering physical, kinetic and thermodynamic properties of the separation.

The equation is related to basically three terms, namely, (A) **eddy** diffusion, (B) **longitudinal** diffusion and (C) **mass** transfer. These properties affect the **broadening** of the peaks in the chromatographic column.

The **van Deemter plot (curve)** is plate height (H) in cm (y-axis) versus linear velocity (v) in cm/s (x-axis).

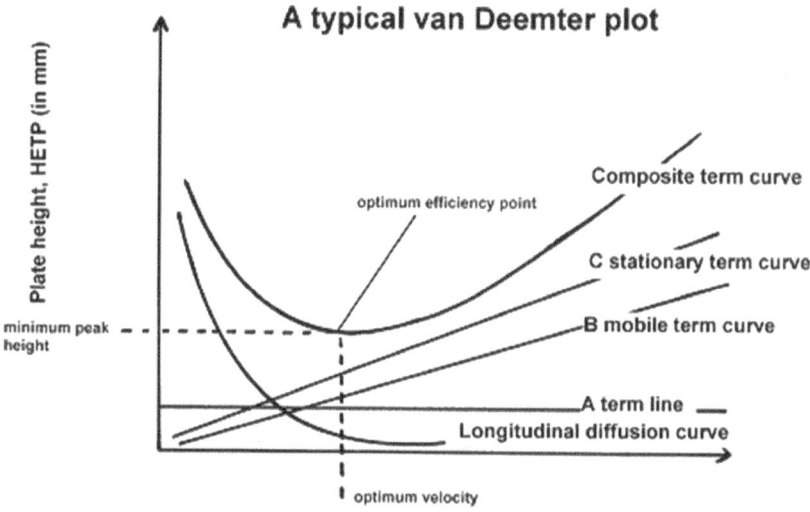

Fig. 6.1: van Deemter plot.

The optimum efficiency of the column is at the base of the u-shaped curve.

$$H = A + B/u + C \times u$$

where v is the linear velocity of the mobile phase (cm/s), H is the height equivalent to theoretical plate (HETP), C is the combination of stationary and mobile flow resistance and u is the average linear velocity.

Note that eddy diffusion results from the unequal flow velocities and path lengths around the packing particles of the stationary phase.

Detector specifications should include:

– **Sensitivity** response (minimum detectable level) with low signal-to-noise ratio (S/N) or signal drift
– A wide **linear** dynamic range
– Precise **selectivity** towards the compounds or substances ought to be separated and measured (peak height or peak area)
– **Repeatability** without introducing any ghost peaks from the previous injection (separation). Sometimes it is necessary to clean or bake-off any residual contaminants on the detector's sensor part

Each type of detector (and different manufacturers of the detector) measures differently the electrical charges or changes that take place during the separation stages.

Kovats retention index is used to convert retention times, in minutes (of an eluting peak in a particular chromatographic column) into a recognisable system of numbers, in order for the chromatographer (analyst) to select the correct column packing stationary phase, for the component of interest being separated.

Tables of these sets of numbers are published and are available on the Internet.

McReynolds classification (constants) is used to compare the polarities of the different stationary phases to that of a standard stationary phase (namely, squalene which is non-polar).

The McReynolds constants and **retention index** (RI) is useful to correctly select a column packing (stationary phase) for a particular type of separation or analysis.

The polarities of the analyte and stationary determine the ability of retention of the analytes (partitions).

To systematically express the retention of an analyte on a given stationary phase (i.e. a GC column) is a measure called RI, which is independent of the column packing, temperatures or any other operating conditions. In other words, RI measures the selectivity of the stationary phase. By definition, RI = 100 × paraffin carbon numbers.

It has five sets of numbers indicating the different degree of polarities of a standard stationary phase with other stationary phases based on the chemical compounds called probes (in order):

– benzene (aromatic);
– n-butanol;
– 2-pentanone;

- nitropropanone;
- pyridine.

Example:
- squalene is 0 0 0 0 0 by definition;
- OV-101 is 17 57 45 67 43; sum = 229;
- Carbowax 20 M is 322,536,368,572,510; sum = 2,308.

Each sum is a measure of the polarity of the stationary phase. The higher the sum, the more polar the stationary phase, hence the higher the retention time.

A technique that is often used in chromatography is **derivatisation**; which converts a chemical compound (e.g. of a sample's functional group) into a similar (more amenable to analysis) product, but of different chemical and physical properties.

Example is **silylation**, which is a process to introduce trimethylsilyl group to make the sample more volatile and thermally stable, as well as easier to separate in the column at lower temperatures.

Another technique is **methylation**; that is the forming of methyl esters of fatty acids when analysing fats by GC, also known as the **FAME** technique. Where FAME is abbreviation for Fatty Acid Methyl Esters.

GC detectors
- **FID** is the most common detector and has an extensive workable linear range (from 1 electrical response unit up to about 10^6 units) of separation of components. It is insensitive to moisture and oxides of gases of carbon, nitrogen, sulphur and silicon. This detector is ideal for **most organic (carbon bonded to hydrogen** functional groups) chemistry analyses.
- **TCD** is commonly used for the analysis of inorganic **gases.**
- **Electron capture detector (ECD)** is ideal for pesticide analysis. The ECD responds to compounds containing **electronegative** functional groups, such as **halogenated** organic compounds. Also used sometimes for highly nitrated and chlorinated organic compounds. The emission source for an ECD detector is a radioactive metal such as ^{63}Ni.
- **Flame photometric detector** is suitable for detecting **sulphur** and **phosphorus** compounds, sometimes halogens. Sometimes, it can also detect arsenic compounds depending upon the GC model.
- **Nitrogen–phosphorus detector** mostly detects nitrogen and phosphorus compounds (and sometimes halogens).

The **units** for minimum detectable quantity are in **grams solute per second**. In other words, it is the amount in grams of the sought analyte components injected per second of the carrier gas flow rate. The linear range is the **maximum amount** that the

Tab. 6.2: GC detector characteristics.

Detector	Minimum detectable quantity	Linear range	Temperature limit (°C)
FID	10^{12}	10^6	400
TCD	10^9	10^4	450
ECD	10^{13}	10^3	350
FPD	10^{11}	10^4	250
NPD	10^{13}	10^5	400

minimum detectable amount can be **increased** in order to still maintain a **direct correlation** between the detector signal response and the concentration of the solute.

Retention time of the chromatogram peak of the solute (analyte component) can be compared directly to the retention time of known substances (standards) to identify the solute.

Response factors are defined as the ratio between the concentration of a compound being analysed to that of the response of the detector to that compound. The response is indicated by a peak on the output of a GC chromatograph, such as:

Response factor = peak area / concentration

There are many factors that can influence the value of this factor, hence for quantitative evaluations of an analyte or compound being tested, it is necessary is to use **relative response factors** and an **internal standard** to calibrate the instrument under its current operating conditions.

Additional types of detectors (such as mass spectrometers) can be added at the end of the eluent column of the GC detector to break down the partitioned components further into smaller groups for identification and quantitation (e.g. GC-MS).

The most **critical part** of a GC (and HPLC) is the **column**. This is where the separation takes place between a stationary phase and a mobile phase based on the partition coefficients of the different components of the sample.

In **principle**, the sample is partitioned between a **mobile** gas phase (carrier gases used are nitrogen, hydrogen or helium) and a thin layer of a non-volatile liquid (e.g. Carbowax), known as the **stationary** phase, coated onto an inert **solid support**, such as silicates, carbon, fluorocarbons or diatomaceous earth.

The selection of the liquid-coated phase is dependent upon the type of sample matrix separation that is required. Various stationary phase characteristics, such as the **Reynolds** constant, can be used to select a suitable phase from the numerous phases available on the market.

The operating parameters of the instrument need to be optimised for each particular type of separation of the sample matrices. These parameters are basically:

- Oven temperature for the column, or **temperature programming**
- **Injector** temperature (e.g. 230 °C); split injection or splitless injection
- Gas **flow rate of the carrier gas** (e.g. the **van Deemter** plot of litres per minute versus HETP)
- **Sample size** (e.g. 1 μL or use of a split sampling valve)
- **Length** of column and internal diameter of the column
- Column **packing** (e.g. SP 2100)
- Mobile **liquid phase**
- **Detector** temperature (e.g. FID at 250 °C)
- On-column injection
- Measurement of either peak **height** or peak **area,** or both (usually this is at the preference of the analyst)

It is important to note that mobile phases or solvents must be degassed before entering the chromatographic column. There are many new and clever ways of doing this; check the Internet.

Evaluation methods
- **Calibration with standards:** quantitative chromatographic analyses involve the normal preparation of a series of standard solutions that are close to the expected composition of the unknown (peaks). Peak heights or peak areas are plotted as a function of concentration versus against detector response (usually in millivolts); this is done automatically by the instrument's software programmes.
- **Internal standard method** is where the internal compound (standard) must be a component that can be completely resolved or separated from the unknown's (sample) peaks. A known amount of this standard is added to the sample (vial), mixed well and injected into the chromatograph. The dilution of the sample caused by the addition must be taken into consideration when calculating the concentration of the unknowns (sample).
- **Area normalisation** is used where the chromatogram represents the entire sample, and where each peak has been completely resolved (separated). The normal procedure is to measure the peak areas, divided by its response factor, to obtain the peak's calculated area. All the peaks' areas (converted to concentrated levels) are then added together and the result should be 100%; if not then a factor is applied to bring the total to 100% composition. Example is the chromatogram of methyl esters of fatty acids of a vegetable oil sample.

6.3 High-performance liquid chromatography

The technique of HPLC (aka liquid chromatography) is basically the same as GLC (aka GC). The only significant difference is that the former does the separations under pressure gradients while the latter does it under temperature (e.g. temperature programming).

The **principle** is the separation of components of a sample by the passage of the sample through a column containing a stationary phase by means of a pressurised flow of a liquid mobile phase. The components migrate through the column at different rates due to different relative affinities for the stationary phase and the mobile phase based on adsorption, molecular size or ionic charge.

An HPLC consists basically of:
- Solvent delivery microprocessor-controlled system
- Columns (normally stainless steel)
- Injection ports or valve injectors
- Syringes, liquid
- Autosampler
- Flow-through detector
- Pumps, pressures are microprocessor controlled

In HPLC, there are generally only **two types of pumps** to create a reproducible pressure supply of mobile phase, to push the solution (liquid) through the column, namely:
- isocratic, which pumps solvent from one reservoir of mobile phase, and
- gradient, which pumps mixed portions of two mobile phases into the chromatographic column.

The **injection system** of the HPLC is important as there are many variations (ports) of loop systems (micro-sampling valves), split and splitless injection.

Detectors for HPLS must have rapid and reproducible responses to presence of solutes, a wide range of linear response to concentration, sensitivity and stability.

Common **detectors** for HPLC are:
- refractive index;
- conductometric for ionic-type samples;
- fluorescence;
- UV/vis detectors or diode array detectors; these can be either of fixed wavelength or variable wavelength; from about 190 to 800 nm; or have fixed wavelengths.

There are other types detectors available depending upon manufacturer and any specific use, such as cannabis/cannabinoids.

HPLC is usually used for the separation of non-volatile substances including ionic and polymeric substances (samples). HPLC is preferable for the analysis of high molecular weight compounds compared to GC.

Ways to **select** the right **HPLC column** for the sample or analyte:
- **Low-molecular-weight** samples (smaller than 5,000 g/mol):
  - Water soluble: ion exchange, reversed phase, hydrophilic interaction liquid chromatography (HILIC) columns.

- Organic soluble: normal phase, reversed phase, non-aqueous size exclusion.
- **High molecular weight** samples (larger than 5,000 g/mol):
  - Water soluble: aqueous size exclusion, ion exchange, hydrophobic interaction.
  - Organic soluble: non-aqueous size exclusion.

The above-listed columns separate the analytes (components or molecules of the sample make-up) based on polarity, affinity, ion charges; also on pore size, bead size and length and internal diameter of the column itself.

The size of the stationary phase **beads** used in the column can affect the backpressure. Smaller size beads yield better resolution but can increase the back pressure.

A smaller **pore size** of the column packing would expose a larger surface area which increases the retention time of the molecular component of the sample (analyte) which alters the separation characteristics. Generally, a pore size of about 100 Å is suitable for smaller compounds of less than 5,000 molecular weights as against the pore size of about 250 Å for larger size compounds of 10,000 molecular weights or more.

A factor to take into consideration is the optimum solvent consumption.

6.4 Ion chromatography

This instrument is basically the same as the HPLC but uses columns that separate by ion-exchange (inorganic compounds, cations and anions) instead of partition solubility properties.

The technique involves an ion-exchange column (similar mechanism to that of the mixed bed resin type water de-ionisers to purify water), and a means of suppressing (removing) ionised species other than that from the sample ions in an eluting mobile phase.

Separation of the analytes is by **cation or anion exchange** resin and eluting (flushing) with sodium hydroxide or similar compound for anionic analytes; or using methane sulphonic acid for cationic analytes.

The eluent from the column passes through a suppressor unit (another column) where any background unwanted electrolytes are removed by converting them to, usually, water or carbon dioxide.

Common stationary phases are:
- Sulphonic acid as a strong cation exchanger, with disodium phosphate as a mobile phase
- Quaternary amines as anion exchanger

Common **detectors** for IC are:
- **Conductivity** detector is the most common as it is used to detect a wide range of analytes such as anions, cations and amines

- **UV/Vis** detector
- There are other types such as electrochemical (amperometry), variable wavelengths, photodiode arrays and mass spectrophotometry

Typical organic applications are vitamins and amino acids separations. Otherwise, the techniques are commonly used in industry to analyse in the ppm ranges for anions, such as nitrates, nitrites, phosphates, halides, sulphates and sulphites, in waters.

6.5 Paper chromatography

The **principle** here is the partition or adsorption (not absorption) chromatography.

In partition paper chromatography (PC), the substances are distributed (aka partitioned) between two liquid phases; water held in the pores of the paper (cf. filter paper) and the other phase being the mobile phase which moves through the paper. This movement of the mobile phase does the separation of the components of the substance, due to the different affinities of the two liquid phases. The amount of movement, that is, separation of the components, is measured by the difference in the travel of the mobile phase from the beginning to the separated component, over the distance travelled by the solvent front. This is the Rf value which is, hence, specific to each separated component of the substance. The Rf value is used to identify the molecular component.

Example is the separation and identification of amino acids (the building blocks of protein molecules).

Adsorption PC is where the paper itself acts as the normal solid phase and the mobile phase is the liquid substance.

The liquid phase of the substance is generally a solution of the sought components of a solid sample and a suitable solvent.

Based on the types of PC depending on the physical setup of chromatography tank and solvent tray, the procedures are:

- Ascending: where the solvent is at the bottom of the tank, then moves in an upward direction by capillary action; this is the most common procedure
- Descending: where the solvent tray is at the top of the tank and the solvent moves downward by capillary action and gravity pull
- Ascending–descending: where the solvent is at the bottom and the paper is looped over a rod at the top, so that solvent moves upwards, then over the rod and then downwards
- Radial and circular: where the sample is spotted at the centre of a circular chromatographic paper and after the sample spot is dried, the solvent is spotted on top of its (the sample) spot
- Two dimensional: this is a very useful procedure where two components could have the same Rf values. The paper is turned 90°, after ascending when the

solvent front reaches about 2–3 cm from top of paper. The ascending procedure is then repeated.

Several factors can affect the Rf values of PC and TLC plates such as
- layer thickness of the coated TLC plate;
- presence of moisture;
- selection of the solvent or solvent blends, as a mobile phase;
- degree of the chromatography tank saturated with solvent vapours;
- temperature;
- presence of sunlight;
- sample size (and circumference of the spot) spotted.

6.6 Thin-layer chromatography

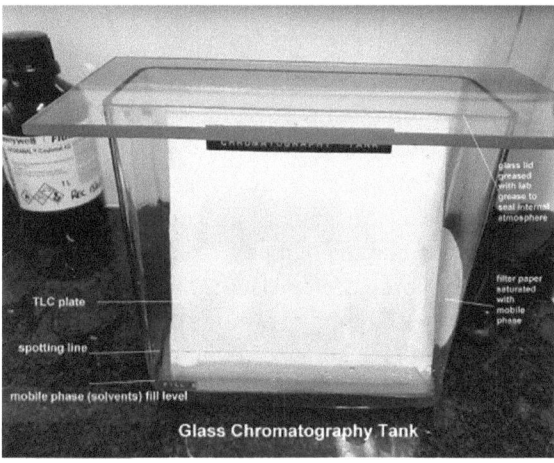

Fig. 6.2: TLC.

The principle here is that a plate is coated with a stationary phase such as silica gel, aluminium oxides, Kieselguhr, polyamides, microcrystalline cellulose and fibrous cellulose. Sometimes the coatings contain a fluorescent dye for UV detection (at 254 nm) of the spots.

The plates are generally made of glass, plastic sheets or aluminium sheets.

TLC is a solid–liquid technique in which the two phases are a solid (stationary such as silica gel 60 coated on glass plate) and a liquid mobile phase, such as a blend of chloroform and methylene chloride. Example of a typical setup is shown in Fig. 6.2 above.

Other common mobile phases are acetonitrile, hexane, chloroform, methanol and isopropanol.

Typical applications are sugar (carbohydrate) profiles, vitamin assays and mycotoxins (e.g. aflatoxins B and G) separations by adsorption.

Example of a method for determining aflatoxins in feed sample:
– Uncap the prepared sample vial, add 0.100 mL (100 μL) of benzene: acetonitrile (98:2 v/v), reseal and shake on vortex machine.
– Draw a pencil line across a pre-*dried* TLC plate at 3 cm from bottom edge of plate.
– In subdued incandescent light and as rapidly as possible, puncture the vial cap with the syringe without bending the needle, and spot 5 μL (two spots) and a 10 μL spot of the sample about 1 cm apart on the pencil line. The spots should be about the same size and not larger than 3 mm in diameter – to aid rapid evaporation of the solvent, it will be necessary to blow air gently onto the spot during spotting.
  Note:
  – Always spot the sample spots first, and then the aflatoxin standard spots – this is to prevent crossover contamination from the syringe.
  – Always drop the liquid on the same centre part of that spot.
  – Rinse out the syringe between samples, standards and so on with the washing solvent of benzene: acetonitrile. After using, rinse the syringe with acetone as acetonitrile attacks the glass syringe.
– On the same plate, spot 1, 3 and 5 μL aflatoxin standards. Also, as an internal control standard, spot a 5 μL sample onto top of another 1 μL standard spot.
– Place the plate immediately into a glass chromatography tank containing the *mobile phase 200 mL (2 cm depth)* chloroform: acetone (180:20 v/v); ensure that the pencil line is not below the level of the chloroform: acetone blend.
– Develop plate 23–25 °C (room temperature) for not more than 90 min (i.e. until solvent front is about 3 cm from top of plate).
– Remove from tank and evaporate solvent at room temperature (about 10 min air-drying).
– Place in UV safety light box and observe the patterns of the four fluorescent spots (beware of UV radiation).
– Examine patterns in the sample chromatograms for the fluorescent spots having R_f value corresponding to and colour appearance similar to those of the standards and identical (if any) with those of standard spots and determine which sample portion matches one of the standard spots (Fig. 6.3).
– *If in doubt*, spray the plate with a confirmatory sulphuric acid to enhance the appearance of the spots, dry the plate and observe the spots again.

The separated components' (coloured spots) concentration can be measured by densitometry or physically removed for further identification.

The components are generally identified by comparison of their Rf values with those of standards chromatographed under identical conditions.

Factor affecting TLC separations (Rf values and resolution), are:
- temperature;

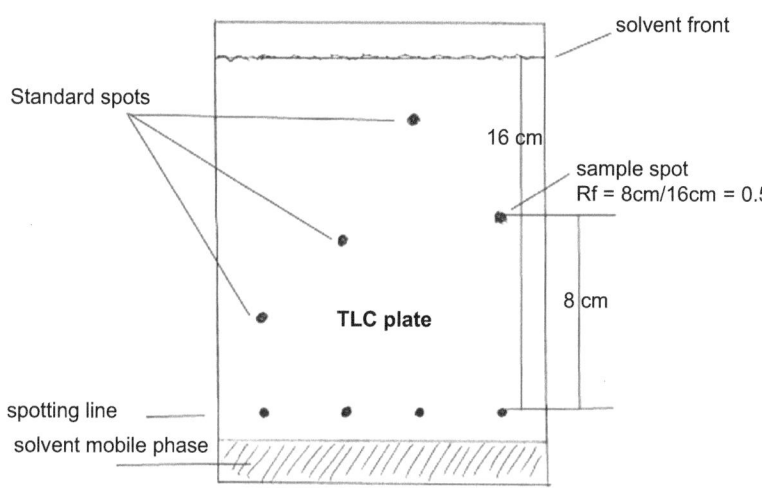

Fig. 6.3: TLC plate.

- time taken for the diffusion process of separation;
- sunlight or UV exposure during separation;
- environment in the chromatography tank; that is, a stable saturated atmosphere of the vapour of the mobile phase (e.g. the liquid or solvent in the base of the tank) is recommended;
- solubilities and polarities of the selected partitioning and/or adsorption systems;
- impurities in the solid phases and imperfections in the uniformity of the coatings;
- overloading of concentration of the spotted area on the base line (that is, sample application and its solvent used for the sample's initial dissolution). Note that the sample solvent and the mobile phase (solvent) are NOT the same.

A more modern, powerful and instrumental technique is high-performance TLC (HPTLC). The main principle difference between the classical TLC and HPTLC is the structure and properties of the separation plate (solid phase). The mean particle size of HPTLC coated plates is about half the size of classical TLC plates (about 6 μm compared to about 12 μm size). These smaller size particles yield a smoother, more compact surface area (higher density), hence improved chromatographic separation.

The benefit of HPTLC is that it produces a faster separation, needs smaller sample spotting and better reproducibility of sharper bands which can be used for further quantitative analysis of the spots or bands.

Examples of the use of this technique are in the fields of nutraceuticals, pesticides, flavonoids and pharmaceutics.

HPTLC has been automated (by use of computerised densitometric measurements) resulting in the widespread use of the technique in quality control laboratories.

A further aspect of HPTLC is the ability of fingerprinting, where the measurement of Rf values, identification and comparisons of the separated bands can be undertaken by the use of modern HPTLC instrumentation. This type of fingerprinting is similar to techniques such as RNA-seq and FTIR, but not identical.

Chapter 7
Conductometry

Keywords: resistivity, specific conductance, absolute conductance, conductivity, units, calibration solutions, K cell constants, TDS, total dissolved solids, salinity, sodicity, conductance values, temperature coefficients, how take measurement

This chapter outlines the principles and procedures of measuring the **conductance of electricity** between two substances (metals); what affects these measurements; and precautions to be taken when handling or cleaning the cells (aka electrodes) and terminology.

An analytical use of electrical conductivity (EC) measurement is in **conductometric titrations** such as neutralisation, displacement or precipitation.

The principle of conductometric titrations is the measurement of the conductance of an electrolyte solution using an alternating current electrical source where the rate of change of conductance is a function of added titrant which is used to determine the end point or equivalence point.

In analytical chemistry, it is the **principle** of measurement of the **EC** in volts (or the movement of electrons, aka electricity, between two (e.g. platinum electrodes, usually 1 cm apart) solutions, gases and solids. An alternating current (AC voltage) is applied across the two electrodes (metal plates) to prevent polarisation at the electrode surfaces.

The conductance of a solution, G, is directly proportional to the electrode surface area, A in cm^2, and inversely proportional to the distance, L, between the electrodes (usually 1 cm).

The constant of proportionality, k in $\mu mho/cm$, is known as the conductivity; such that:

$$G = k(A/L)$$

Reversal of conductivity is the study of the resistance to the conductance of electrical currents in various types of matter. **Conductance**, G, is defined as the reciprocal of resistance, R:

$$G = 1/R$$

where R is in ohm; G is in ohm^{-1} or mho, or Siemen S.

Equivalent conductivity of a solution or liquid is the conductivity per unit of concentration.

Resistivity, measured in ohms (Ω), is the inverse of conductivity, measured in Siemens (S).

Specific conductance (by definition) = cell constant × conductivity

https://doi.org/10.1515/9783110721201-008

where cell constant, K is the *specific conductance at t °C* and EC at t °C.

Conductance readings × cell constant = conductivity units (this is usually calculated automatically by the meter).

It is the **conductivity**, in microSiemens, at a certain temperature (20 or 25 °C) that is usually recorded in laboratory reports.

Absolute conductance (by definition) = conductivity units

Note that when temperature increases, so does its conductivity, as more energy is given to the electrons to flow faster (mobility), at a **rate** of approximately **2% per degree centigrade**, but the rate actually depends upon other various factors, such as:

Nernst – Einstein equation: $EC = D$

and

Stokes – Einstein equation: $D = \eta$

Conductivity is determined from the voltage and current values according to **Ohm's law**:

Conductivity (current) = voltage/resistance

where current (I) is in amps, voltage (V) in volts and resistance (R) in ohms.

The ions in the solution (sample) causes a current to flow under the applied voltage; hence, the conductivity (or passage of ions moving from one electrode plate to the other plate) in a solution or solid is proportional to the sample's ion concentration.

In most dilute aqueous samples, the relationship is linear, but in others, the ionic interactions of high conductivity (e.g. high salt or mineral content) can lead to a nonlinear relationship between conductivity and concentration. This can be somewhat corrected by using a cell with a higher K value.

The electrode plates are usually made of platinum and are 1 cm apart (hence, measuring units are per cm). The two electrodes are combined into a single unit to form the **conductivity cell**.

The **EC** measurement uses a calibrated conductivity meter (which is basically an ohm or resistance meter). The cell K value is **calibrated** against known concentrations of **potassium chloride** solutions.

Calibration solutions

Standard KCl calibration conductivity solution:

Note: 0.0100 M KCl = 1,412 µS/cm at 25 °C or c (KCl) = 0.0100 mol/L = 0.0100 N KCl

For 1.0 M, take 0.7456 g (dried at 110 °C) potassium chloride (KCl) Analar® Grade, dissolved in water and made up to 1 L volume; store solution in plastic bottle at 4 °C.

Alternatively, prepare as below for the closest match to that of the sample to be tested:

KCl concentration in mol/L		Conductivity equivalent in μS/cm
0.0001	=	14.94
0.001	=	146.9
0.01	=	1412
0.1	=	12,890
1.0 (74.56 gKCl)	=	111,900

Cell constants (K) of the conductivity cell, should be:
- for samples of very low conductivity, range <1 μS (e.g. purified water), K = 0.001;
- for samples of low conductivity, range 1–20 μS (e.g. lab grade water, boiler feed water), K = 0.01;
- for samples of conductivity, range 20–400 μS (e.g. municipal and potable waters), K = 0.1;
- for samples of moderate conductivity, range 400–1,000 μS (1 mS) (e.g. hard water, grey (gray) waters and trade effluents), K = 1.0;
- for samples high in conductivity, range 1–200 mS (e.g. brine, seawater), K = 10.0.

Total dissolved solids

The conductivity measurement method is commonly used to measure the total dissolved solids (TDS) of various waters to indicate properties such as hardness, scaling or corrosion.

Conductivity of water may be considered as the degree of mineralisation (inorganic matter or electrolytes) present in the water.

The correlation between TDS and EC can be stated by a simple equation (at 25 °C):

$$TDS = \alpha \cdot EC$$

where α is an **empirical value** for a particular type or matrix of water sample.

To mathematically convert conductivity units to TDS units:

$$TDS \text{ mg/L (or TDS ppm)} = \text{conductivity } \mu\text{S/cm} \times \text{empirical factor } \alpha.$$

where empirical factor α can be any number from 0.55 to 0.9, depending upon the soluble components of the water sample (and temperature).

Example:

For boiler waters: factor at 20 °C = 0.70 or at 25 °C = 0.63

For irrigation and natural waters: factor at 25 °C = 0.65

For other waters: factor at 25 °C = 0.75

Note: Meters that automatically display TDS units are set on the manufacturer's own empirical factor! This could be anything from 0.4 to 1.0.

This factor or value α is basically a ratio of EC:TDS.

However, on some instrument models, the "conductivity-to-TDS factor" can be adjusted by the chemist on the instrument's menu or set-up page.

Some conductivity meters will also, in addition, display salinity readings.

For boiler waters and on all alkaline waters, the phenolphthalein alkalinity must be neutralised to pH 7.0 ± 0.2 by addition of a trace amount of 1% acetic acid before doing conductivity measurements; this is to overcome the high conductivity of the hydroxide ion in boiler waters.

Salinity:

To convert conductivity units to salinity units, use is made of a complex equation:

Salinity in g/L = (EC in mS/cm to the power of 1.0878) × 0.4665

Salinity of seawater is about 35 ppt or 35 g/L (i.e. about 3.5%, but it varies from ocean to ocean).

Sodicity:

For soils where a 10% aqueous mixture is made, the EC is referred to as sodicity.

Note that sodicity is the measurement of conductivity of soils and expressed as deciSiemen per metre (dS/m); some of these meters can also convert conductivity to **moisture concentration.**

Soil conductivity, also known as **saturation extract EC**, is measured by making a paste with the soil sample and deionised water, filtering and measuring the conductivity of the filtrate.

A quicker procedure is to create a 10% mixture of the soil and carefully measuring its conductivity without damaging the cell Pt plates.

Conductance:

Calculations can also be made about the concentrations of dissolved electrolytes from the conductivity when the ionic composition and equivalent conductivities (valency of the ion) or conductance of the sample components are known. Hence, the conductivity can be calculated instead of measurement by a meter. In other words, the EC of inorganic aqueous liquids can be calculated from the conductance values of its cations and anions, if the complete ionic chemical composition of the liquid or sample, is known (Tab. 7.1).

Tab. 7.1: Conductance values equivalent conductivities at infinite dilution at 25 °C in S cm^2/mol.

Cation	C$^+$
H$^+$	350
Na$^+$	50
K$^+$	74

Tab. 7.1 (continued)

Cation	C^+
Li^+	39
NH_4^+	74
Ag^+	62
Cs^+	77
Sr^{2+}	119
Ca^{2+}	119
Ba^{2+}	128
Mg^{2+}	106
Zn^{2+}	106
Cu^{2+}	107
Pb^{2+}	142
Ni^{2+}	106
Ba^{2+}	127
Fe^{2+}	108
Fe^{3+}	204
La^{3+}	210
Al^{3+}	183

Anion	A^-
NO_2^-	72
OH^-	199
F^-	55
Cl^-	76
Br^-	78
I^-	77
IO_3^-	41
IO_4^-	55
ClO_4^-	67
ClO_3^-	65

Tab. 7.1 (continued)

Anion	A⁻
BrO_3^-	56
$HCOO^-$	55
HCO_3^-	45
CO_3^-	139
SO_4^-	160
PO_4^-	207

When a sample conductivity is measured at a temperature other than 25.0 °C without automatic internal meter temperature compensation available, then the EC can be converted to the standard reported temperature of 25.0 °C (or 20.0 °C).

This can usually be done by calculation.

However, the temperature coefficient used to adjust the waters to a given temperatures varies depending on the kinds and concentrations of electrolytes (ions) present, temperature at time of measurement and the temperature to which measurements are being adjusted (e.g. 25.0 °C).

The **temperature coefficient**, c, reflects the rate of change of conductance per degree of temperature change.

This figure is usually taken between 2% and 2.5% (used to be 1.9% on old instrumentation) per degree centigrade at 25 °C (c = 0.025) for meters that display the conductivity reading at a set temperature of 25 °C.

To calculate the conductivity at the measured temperature of the sample, apply the following equation:

$$\text{measured conductivity in } \mu S/cm \text{ at } t\,°C = \text{reading} + 0.025\,(t\,°C - 25\,°C)$$
$$\text{for } t\,°C \text{ above } 25\,°C$$
$$\text{measured conductivity in } \mu S/cm \text{ at } t\,°C = \text{reading} - 0.025\,(t\,°C - 25\,°C)$$
$$\text{for } t\,°C \text{ below } 25\,°C$$

where: t °C = measured temperature of sample reading = current reading or display on meter

Another way is when the sample conductivity is measured at its own temperature on a meter that is without internal temperature compensation, then conductivity at 25 °C can be calculated by the following equation:

$$\mu S/cm \text{ at } 25\,°C = \frac{\text{measured conductance in } \mu S/cm \text{ at } t\,°C}{1 + 0.019(t\,°C - 25\,°C)}$$

where t °C = measured temperature of sample.

Units of measurement

SI system uses mS/m (milliSiemens per metre)

Note: 1 µS/cm = 1 µmhos/cm

1 mS/m = (µS/cm)/10 = 10 µmhos/cm

1 mS/cm = 1,000 µS/cm

1 dS/m = 1 mmho/cm

1 dS/m = 1 mmho/cm

1 dS/m = mS/cm

where S is Siemens = mhos; mho is the "inverse" of the unit of resistance, namely ohm.

How to take correct conductivity measurements?

Example:

Apparatus:

Conductivity meter (with automatic temperature compensation) having a known cell constant for its electrode (cell); and calibrated using a standard reference KCl solution (see below).

Keep cell *clean* (*do not* touch the black film on the internal platinum plates).

For organic dirt, immerse cell in hydrogen peroxide or sodium hypochlorite (bleach) or alcohol for about 1 h, then flush well with distilled water.

For inorganic or mineral dirt, immerse cell in dilute EDTA solution or dilute hydrochloric acid (<1 M), for about 1 h, then flush well with distilled water.

To clean very dirty cells, flush very quickly and carefully in very dilute chromic acid.

As a last resort, to re-platinise (i.e. if the platinum-black has flaked) refer to instrument manufacturer's manual.

Note: Always keep the cell *dry* when not in use (to prevent algae build up), then soak in water for about 1 h before use.

Thermometer, if instrument, has no automatic temperature compensation.

Select a suitable cell (refer to cell constant K) to cover the expected conductivity range of sample, e.g. K = 1.0000.

Reagents:

Distilled or de-ionised water of conductivity is less than 1 µS/cm, for preparing calibration standards, diluting samples and rinsing cells.

Procedure:

Operate the meter as per the manufacturer's instructions. Make sure that there are no trapped air bubbles between the plates in the cell and the sample liquid; the cell should be immersed in the sample to at least 10 mm above the open aperture in the stem of the cell.

For boiler waters and all hydroxide alkaline waters, the phenolphthalein alkalinity must be neutralised by the addition of a drop of 1% acetic acid before doing conductivity measurements – this is to overcome the high conductivity of the hydroxide ion.

If the sample temperature is different from ambient temperature, then allow the cell to equalise for a few minutes in the sample before noting the meter reading.

Calculation:

If the cell constant, K, is not exactly 1.0000 and/or the meter cannot be set on a known cell constant value, then all meter readings are in conductance units; hence, conductance readings × cell constant = conductivity units.

The instrument should automatically compute the correct conductivity.

Report:

Report all conductivities at temperature of 25.0 or 20.0 °C, to not more than three significant figures.

Some typical quality parameters for waters methods, in general, are:

Standard deviation is about 8%.

Precision is not better than 1%.

Reproducibility is not better than 2%.

Chapter 8
Coulometric analysis

Keywords: Karl Fischer, coulombs, Hydranal reagents, Faraday's laws

Coulometry is a group of methods which involves measuring the quantity of electricity (in coulombs) required to convert the analyte quantitively to a different oxidation state.

The **principle** of this technique is basically the measurement of the quantity of electricity that is used in an electrochemical reaction at constant potential (voltage) or constant current (amperage).

The advantage of this technique is that the standardisation step is not necessary as the coulombs measure is proportional to the weight of analyte.

However, when water is determined by this method (e.g. Karl Fischer (KF)), a water standard or salt of known water of crystallisation is required to check the strength of the coulomat reagents used.

It is the study and practice of changes in electric charges; the most common instrument is the KF moisture analyser (coulometric model, not the titrimetric volumetric model which is more costly to run as it requires an extra chemical reagent). Example of this type instrument is shown below in Fig. 8.1.

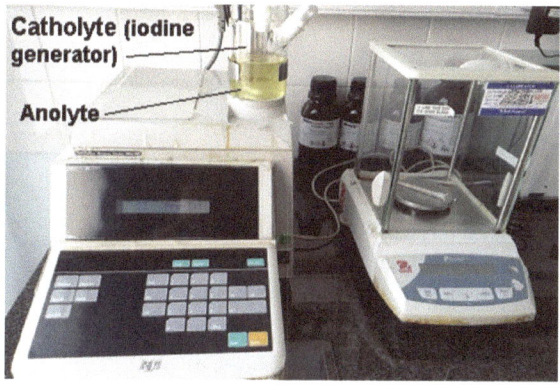

Fig. 8.1: Karl Fisher moisture titrator.

The **principle** of this type of titration (opposed to the common KF volumetric titration) is that the **iodine is created by electrolysis** from a special coulometric KF reagent, which is a mixture of KF reagent and a solvent in which the sample is introduced. An electrical current is applied to an electrode until the end point of the titration is reached:

$$H_2O + SO_2 + I_2 = SO_3 + 2HI$$

https://doi.org/10.1515/9783110721201-009

1 mol H₂O (18 g) = 2 × 96,500 coulombs = 10.72 coulombs per milligram H₂O

The instrument and reagents should be checked and calibrated on a water standard (Hydranal® – water standard 0.10) or pure distilled water, before use. Example is to inject 1 µL pure water into the titration vessel and check if instrument displays a water content of 100%.

The **coulometry principle** in the KF technique is basically where the amount of a substance is transferred electrically from something (usually the iodine generator glass cell or catholyte cathode) to something else (another electrode, the detector platinum electrode) by the measurement of a voltage impedance or current change. The solvent/solution acts as the anolyte.

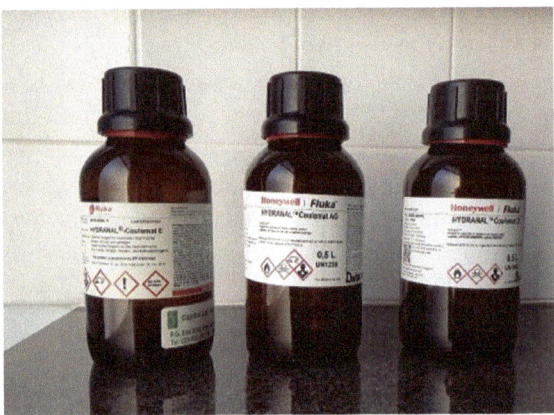

Fig. 8.2: Hydranal reagents.

Examples of common reagents used are (Fig. 8.2):
- Hydranal® Coulomat E #34726 the anolyte reagent based on ethanol as solvent (flammable).
- Hydranal® Coulomat AG #34836 the anolyte reagent based on methanol as solvent (flammable).
- Hydranal® Coulomat Oil which contains solubilisers, useful for petroleum oils, e.g. ASTM D4928; but also suitable for silicone oils and vegetable oils.

Other types depending on sample matrix and instrument (diaphragm or diaphragmless) are:

AD, AK and AG-H.

The anolyte-containing reagent is only required for a coulometric KF titration, as the catholyte is self-generated inside the instrument's glass cell unit.

The coulometric procedure has a lower detection limit (about 10 μg water) than the volumetric procedure and can utilise lower amounts of sample than the general volumetric method.

The most **common error** is **air/moisture** seeping into the titrating unit through loose fittings or ungreased joints. This leakage problem causes a drift in the signal display, thus making it difficult to discern when the titration has ended. Some instruments automatically compensate for this drift if it is at a constant or stable rate.

The desiccant tube on the instrument must be re-generated (dried) frequently.

Another cause of errors or non-completion of a titration is the build-up of non-polar substances (from a sample's chemical composition) causing a lowering of the conductivity of the reagents.

The sample size injected into the titrating vessel should be about 0.5 mL to maximum amount of 5.0 mL depending upon the expected level of moisture present in the sample. For solid samples, the mass will be the equivalence to the liquid size.

The capacity limit of the Hydranal anolyte is seldom reached before the titrating vessel needs to be emptied. Thus, the advantage of this technique is that samples can be added repeatedly to the cell, after each titration, until the vessel becomes full.

The theory and practice of coulometry are based somewhat on Faraday's law of electrical changes.

Faraday's law states that the amount of substance deposited on an electrode is proportional to the total electrical charge (volts) that is transmitted through the electrolyte and the amounts of different substances liberated at an electrode by the same current (amps) flowing for the same time period is proportional to their chemical equivalents.

Current in amperes multiplied by time in seconds is known as a **coulomb** unit.

Note that the KF method only reacts with water (**free** water, emulsified samples and **bound only if** the samples dissolves in the KF reagent); not like other volatiles which are determined in the common oven-dried moisture method.

Solvents used in the KF reagent are usually ethanol or methanol.

Chapter 9
Potentiometry

Keywords: ion-selective measurements, pH, causes inaccurate measurements, potentiometric titrations, buffer capacity, how to take pH measurement, factors affecting measurements

Ion-selective measurements

The **principle** here is the measurement of a potential difference between two electrodes (a reference and the sensing electrode) caused by the migration of ions through a membrane from a high concentration of ions to a low concentration of ions. The electrodes are connected to the meter. The pH electrode has both the reference and indicator electrodes encased into a single junction unit. A temperature-sensing device is also connected to the meter.

The measurement is done in millivolts; hence, the meter itself is essentially a voltmeter.

There are four main types of electrode materials and membranes:
- glass membranes, e.g. H^+ used for pH measurements;
- solid-state sensors (electrodes) or crystalline type, e.g. F^-;
- polymer or liquid based, e.g. Ca^{2+};
- compound electrode, e.g. CO_2 gas-sensing electrode.

The solid-state sensors have a sensitive membrane made from compounds of the metal (ion) being measured, e.g. the fluoride membrane is made from lanthanum fluoride and other compounds to complete the electrochemical cell system.

These types of measurements are subject to many errors and interferences. Chemical interference is when the sample matrix blocks the membrane from sensing the ion of interest and electrode interferences when the membrane senses ions other than or in addition to the ion of interest.

The most reliable one (other than pH or H^+) is the fluoride (F^-) electrode. Other common electrodes are calcium and ammonium.

Note that the electrode should only respond to the activity of the **free ion** (analyte) in solution.

Common causes of inaccurate readings:
- Selectivity is poor, e.g. Na^+ correction for the pH measurements; basically, interference of other ions present.
- Precision is never better than 2%.
- Unstable meter readings: this problem could be smoothed out by the electronics (capacitors) of the meter, to give the false impression that the electrodes are clean and stable.

https://doi.org/10.1515/9783110721201-010

- pH effect: for example, the fluoride readings which should not be taken at high pH levels (above 8) due to the hydroxide interference; preferably, sample pH should be between 5.2 and 8.3 for fluoride measurements.
- Dirty or blocked membranes.
- Fragile electrodes.
- Electrolyte can become contaminated by infusion of a sample's chemical makeup or matrix.
- Build-up of electrostatic charges on the outer casing of the electrode: this can be prevented by earthing the electrode with some copper wire attached to the electrode casing (if any, as most modern instruments are now made with plastic casings) and then connected to an earth or metal water pipe.
- Ion-selective electrode) measurements (e.g. fluorides) can be plotted manually or by using Excel spreadsheets on semi-log graph paper where the ordinate scale (y-axis) is the logarithmic scale of the concentration of analyte (e.g. mg/L F^-), and the abscissa (x) is the common unit scale of mV.

Potentiometric titrations

The **principle** is based on the **Nernst equation.** This is an equation that relates the reduction potential of an electrochemical reaction (half-cell) to the standard electrode potential of the chemical species that are undergoing reduction and oxidation.

On automatic titrators, the end point of the reaction is displayed automatically; otherwise, the analyst can plot (or via the instrument's software) the inflection point of the curve graphically.

More precise determination of the end point (equivalence point) is to measure the rate of millivolt (cell emf, pH) change (the first derivative) against the amount of titrant added, to determine the endpoint of the titration. Determining the second derivative will yield a more precise value of the endpoint.

Basically, an electrode is used as an indicator (against a reference electrode) to determine the endpoint of the titration in place of chemical indicators.

The other advantage of potentiometric titrations over the usual volumetric titration is that the multiple species of the analyte can be determined simultaneously in the same titration. Example is the determination of the individual acids present in wine, namely lactic, malic, citric and tartaric acids in the sample; this thus yields different end points during the same titration.

pH measurements

What is **pH**?

It is a measured amount of acidity or alkalinity (basicity) of a substance.

An electrometric method (potentiometric measurement) is commonly used to determine this degree of acidity or alkalinity of substances, i.e. pH meter.

Temperature greatly influences the pH readings because of the **Nernst equation** for electrometric measurements.

Note: pH (by definition) is $-\log_{10}[H^+]$

i.e.

$$pH = \frac{1}{\log_{10}[H^+]}$$

where $[H^+]$ is concentration of the hydrogen ion, in moles per litre; p denotes the logarithm function of numbers to a base 10.

Example: a pH reading of 6.6 of a liquid would indicate a concentration of $10^{-6.6}$ g hydrogen ion (H^+) per litre sample solution.

The **pH scale** is an arbitrary scale such as: $pH + pOH = pK_w = 14$

where p denotes the negative log function and pK_w is the negative log **ionisation constant** for water at 25 °C.

Thus, a pH 7.0 at 25 °C denotes a neutral water (i.e. it is $[H^+] = 10^{-7.0}$ mol/L).

Examples of some ions in water:

- below pH 2 denotes HSO_4^- or free acids present, or
- below pH 10 denotes S^{2-}, SO_3^{2-}, PO_4^{3-}, AsO_4^{3-}, AsO_3^{3-}, CO^- or $B(OH)_4^-$ present.

Buffer capacity is where a solution that contains conjugate acid–base pair, possesses remarkable resistance to changes in pH. The buffering capacity of a solution is defined as the number of equivalents of strong acid or base needed to cause 1.00 L litre of the buffer to undergo a 1.00 unit change in pH.

How to take a **correct pH measurement?**

1. *Apparatus*:

- A pH meter capable of reading to 0.01 pH units with a temperature compensator. Meters are available that can display readings to nearest 0.001 pH units, but these are only of any practical use for research work where comparisons or monitoring reactions are required.
- Best practice is to standardise meter before every use, using at least three fresh pH buffer solutions, one pH 7 (neutral point) and others are pH 10 and pH 4.
- A thermometer if instrument has no automatic temperature compensation.
- A combined glass electrode (or any other suitable electrode system for the type of samples to be tested). If applicable, use only the refillable-type electrodes for best results. The ideal slope for a good pH electrode is between 95% and 110% – some meters will display the slope value immediately after calibration.
- Check periodically the level of electrolyte in the electrode and refill if necessary with electrolyte solution – make sure the correct strength and composition of the electrolyte is used for that particular manufacturer's making, e.g. 3 M KCl, 3.3 M or saturated KCl solution (about 4 M).
- **Handle the electrode with great care. Store electrode in saturated KCl solution or electrolyte storage solution.**

Do not use the electrode as a stirrer!

2. *Reagents*:

Distilled or de-ionised water of conductivity less than 5 µS/cm for preparing buffer solutions, diluting samples and rinsing electrodes.

pH buffer solutions are commercially available from many suppliers. Most common range is pH 4 (acid range), pH 7 (neutral point) and pH 10 (alkaline range). pH buffer solutions have a short shelf life due to exposure to the air (can be extended somewhat by storing in refrigerator) and should be stored in a cool place and away from direct sunlight. Discard if there is any presence of algal growth or change in colour.

3. *Procedure*:

Operate the meter as per the manufacturer's instructions. Make sure that there are no trapped air bubbles adhering to the electrode.

Uncover the cap and fill hole in the electrode to equalise pressures.

Select two buffer solutions (for **calibrating** the meter) that will bracket the expected pH of the sample; always first set the isopotential point (0 mV) with the pH 7 buffer solution.

For accurate measurements bring the temperature of the sample to the **same temperature** as that of the buffers used to calibrate the meter, or vice versa. **(Note that the temperature compensating device of the meter only corrects for changes in temperature of the buffer solutions and not of the sample).**

Always allow the electrode to equalise temperature if sample and calibration buffer are at different temperatures. Preferably the sample's temperature should be adjusted to same temperature as that of the calibration.

Swirl or stir the sample gently (without forming a vortex) whilst taking a pH reading (to avoid localised ionisation on the electrode) and note the final stable reading after waiting for temperature equilibrium (at least 1 min, but not longer than 3 min).

Always rinse electrode between readings with distilled water; avoid excessive wiping of the electrode which could cause electrostatic charges to accumulate on the electrode casing.

If readings show slow response, drift and poor slope value or are unstable, then the electrode is dirty or faulty. Refer to separate section below for problem solving (note that a pH meter itself is essentially a simple millivolt meter and hence it (the meter not the electrode) should never require servicing or repairs).

4. *Report*:

A theoretical temperature correction factor that may be used to convert a pH reading to a different temperature is:

0.003pH units/°C/pH unit difference from pH 7 calibration point

For more accurate temperature compensations, a graph should be compiled of pH readings of the specific type of sample matrix versus temperature.

Always report the temperature (standard temperature is 25 °C) at which the pH was measured, and report the pH to not more than two decimal places (for very pure waters where a constant reading cannot be determined due to lack of electrolytes in the water, report the pH to the nearest 0.1 pH unit).

5. *Precision/tolerance*:
Standard deviation is about 0.02 pH units. Note that the pH result of a sample can never be better or more accurate than that of the buffer solutions used in the calibration and the quality of the electrode.

Also, electrode errors could occur at below pH 2 and above pH 12, depending upon the construction of the electrode, especially due to high concentrations of the sodium ion (Na^+).

In summary, errors affecting pH measurements (with the glass electrode) are:
Alkaline error due to presence of alkali–metal ions (such as Na^+) in the sample liquid at pH values greater than 10;
- acid error at values below pH 1;
- dirty surface coatings of the electrode;
- blocked membranes or diaphragms;
- dehydration or drying out of the internal electrolyte solution;
- blockage of the filling hole with crystallisation of potassium chloride salt, thus causing an imbalance of pressure differentials;
- refilling or topping up the electrolyte with the incorrect strength potassium chloride solution and formulation;
- attempting to measure pH levels of purely buffered neutral liquids, such as ultra-pure water;
- errors in preparation of the standard buffer solutions for calibrations of the pH meter, or expired buffer solutions.

Note that ultra-pure water (<1 μmho/cm) cannot conduct electricity as there are no mineral ions present. To attempt to measure the pH of such pure water, usually a small drop of pH 7.0 buffer solution is added, in order to add some phosphate ions to enhance the flow of electricity.

Critical maintenance of a pH meter and its electrodes is critical for accurate measurements.

The most common fault of all pH meters is dirty or blocked electrodes. The meter itself (which is basically a millivolt measuring meter) is very robust and requires no maintenance, other than to be kept clean and free of liquid splashes or corrosion.

There are many types of **pH electrodes** available and it is best to purchase one that is robust, accurate **and** designed for use with the particular type of samples/liquids to be measured.

Some problems experienced with pH electrodes (e.g. slow response, drift, unstable readings and poor slope values) are due to:
- Air bubbles in the filling chamber; tap the glass wall of the electrode gently to dislodge the bubbles.

- Blocked reference junction/diaphragm (brown AgCl) and bulb contamination with insoluble particulate matter or proteins. Clean as indicated below.
- Crystal formation (KCl salt crystals) inside junction or main body of electrode (due to leaving the filling hole open when not in use) may be re-dissolved by heating to about 50 °C under running hot water.
- Trying to determine the pH of a non-aqueous solution/sample (always use a special pH electrode from supplier for measuring pH in organic solvents and non-aqueous solutions).

Basically, by definition of pH, there must be present in the sample to be measured, traces of either hydrogen ions or hydroxides ions; in other words, water!

Dirty electrodes can be cleaned using various procedures depending upon what type of samples/liquids were tested.

For example:

- "clean" samples, such as water samples, just rinse under hot running water;
- samples of high proteins such as blood, milk and so on, dip into a mixture of 0.1 M hydrochloric acid and 5% pepsin, for a few minutes only (but not longer than 30 min); or weakly acidic 7% thiourea;
- samples of high organic matter such as foodstuffs, trade effluents and oils, dip into 50% acetone or alcohol for a few minutes only;
- replace the electrode solution with fresh electrode filling solution; do not remove any internal black crystals (AgCl);
- for very dirty electrodes, soak in 0.1 M ammonium bifluoride solution for maximum 2 min; and
- as a last resort, take a very fine "nail file" (or toothpaste) and gently scrap away any black substances on top layer of the membrane diaphragm.

Note that after any of the above remedies, always wash electrode thoroughly in clean running water, and then let stand overnight in an "electrode storage solution", or pH 7 buffer solution.

It is also recommended that on a regular basis the internal electrolyte (solution) should be totally replaced by fresh electrolyte (this is not applicable to gel-type electrodes).

Electrodes should not be stored dry when in regular use; rather store in an "electrode storage solution" available from laboratory suppliers, or its electrolyte filling solution (e.g. 3 M KCl); or pH 7 buffer solution for short periods.

Note: Always top-up or refill the electrode with the correct strength electrolyte (refilling solution)! If you are not sure, then check with your electrode supplier.

There are various **strengths of potassium chloride solutions** that are used as electrolytes such as 3 M (3 M KCl), 3.3 M and 4 M (saturated); also, some electrolyte refilling solutions are saturated with silver chloride (AgCl).

Chapter 10
Polarimetry

Keywords: Biot's law, polarisation, terminology, scales, specific rotations, polarimeters, pol, saccharimeters, polariscope, quadrant, angular degrees, dextrorotatory, laevorotatory

The basic **principle** here is the measurement of the degree of rotation of polarised light, when passed through a substance that has the property of changing the angular rotation of the polarised light beam to either the left (−) or to the right (+).

Polarisation is a process where a source of monochromatic light (the sodium D line at a wavelength of 589.3 nm) is passed through a solution of an optically active substance (such as sucrose) which rotates the plane of polarisation; here the degree of rotation is dependent on the amount of the substance in solution.

The angle of rotation and concentration of the liquid have a **linear relationship**, up to a concentration of about 3 optical density (OD) units.

Temperature is critical and is usually measured at 20.0 °C (but sometimes can be at 25.0 °C).

Polarimeters are typically used for measuring **concentrations** of sugar solutions, isomers, peptides and volatile oils. Saccharimeters are polarimeters used in the sugar and confectionary industry.

If sucrose is the only optically active substance present in the sample, then a 26% aqueous solution at 20 °C in a 200 mm pol tube will give a reading directly in %sucrose in sample.

The wavelength of the **light source** is critical (normally at **589.3 nm**) for accurate readings; the various other wavelengths used in industry are 365, 405, 436, 546 and 578 nm.

In general, the observed optical rotation at 436 nm is about double and at 365 nm about three times that at 589 nm.

A manual traditional polarimeter was known as a polariscope. A scope is a round tube through which one observes or measures something; examples are telescope, endoscope and rifle scope.

Hence, the notation of quadrants of a circle when displaying the angles of polarised light.

A plane of polarised light being rotated **clockwise** (i.e. to the right, or with a+ notation) is referred to as **dextrorotatory**; a plane of polarised light being rotated **anti-clockwise** (i.e. to the left, or with a− notation) is referred to as **laevorotatory**.

Depending upon the angular range of the instrument, one must be diligent when reporting the readout as all polarimeters can only measure the observed rotation of a plane of polarised light.

https://doi.org/10.1515/9783110721201-011

Vertical line is the plane of light.

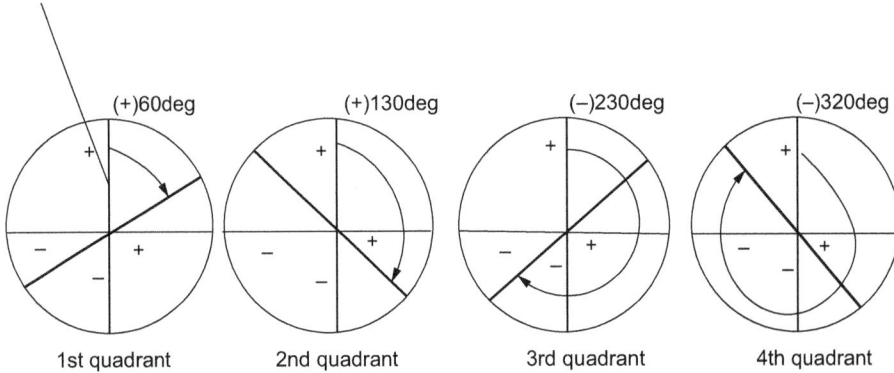

| (+)60deg | (+)130deg | (−)230deg | (−)320deg |

1st quadrant   2nd quadrant   3rd quadrant   4th quadrant

Fig. 10.1: Quadrant.

Note that in the second quadrant of Fig. 10.1, the sample has rotated the plane 130°. That is, dextrorotatory or (+)130°.

In the third quadrant of Fig. 10.1, the sample has rotated the plane now to 230°, dextrorotatory. Hence, the instrument could show an observed rotation of −130° (i.e. a laevorotation), instead of the sample's true rotation of 230°.

Note that in the fourth quadrant of Fig. 10.1, the sample has rotated the plane now to 320°. Hence, the instrument could show an observed rotation of −60° (i.e. a laevorotation), instead of the sample's true rotation of 320°.

Biot's law

The law states that an optically active substance (e.g. sugar solution) rotates plane-polarised light through an angle inversely proportional to its wavelength (e.g. sodium D line at 589 nm).

A 20 W tungsten halogen lamp may be used as a stable white light in conjunction with applicable narrow band interference filters to select the required light source wavelength.

Optical rotation is also affected by the **solvent used** (usually water) for the measurement, and thus this should always be specified.

Other **dependent variables** on the rotation are:
- **Type** of sample, such as liquid sugars, essential oils, peptides, volatile oils
- **Concentration** of the optical active component of the sample
- **Length** sample (pol) **tube** (length of beam light passing through the sample liquid)
- **Wavelength** of the **light source** (Tab. 10.1)
- Type of **detector**, generally a photodiode, capable of measuring amount light of up to 3.0 OD

- **Temperature** of the sample liquid inside the pol tube or the compartment and can be controlled by a Peltier temperature controller (Tab. 10.2)
- **Entrapped air bubbles** in the pol tube during filling of sample liquid
- Dependent upon model or type of polarimeter, resolution of measurement of rotation can be as good as 0.001 angular degrees (Å) with an accuracy of ±0.010 Å; reproducibility or stability can be better than 0.002 Å
- Over-tightening the quartz (deck) glass fastening rings that are at each end of the pol tube can cause invisible stresses in the glass (quartz) structure, leading to skewed angle measurements

Tab. 10.1: Specific rotation sucrose at different light sources.

Light source	Specific rotation
Sodium yellow light at 589.3 nm (industry standard)	+66.5°
Yellow LED (plus interference filter, 589 nm)	+66.5°
Mercury green light at 546.2 nm	+78.4°
Laser light at 633.0 nm	+57.2°
Near infrared at 882.6 nm	+28.5°

Note that the **unpolarised** light from the light source is first **polarised** by passing it through a **polariser** (e.g. Nicol prism), then through the **optically active substance** that is in the long (usually 200 mm) pol tube and rotated plane of the polarised emitted light is measured by the **analyser** (e.g. Nicol prism) and is finally captured and displayed by the polarimeter's microprocessor.

Also, note that the type light source of an instrument depends upon the manufacturer and model of the polarimeter.

Some polarimeters offer flow-through cells or pol tubes, for online processes.

Some manufacturers offer both refractive index and optical rotation on the same model.

Tab. 10.2: Rotation of a sucrose solution at different temperatures.

Temperature (°C)	Rotation
20	+40.00°
21	+39.98°
23	+39.94°
25	+39.91°

Temperature correction to another temperature, say T °C, for the specific rotation of a sugar solution can be calculated by using the following equation:

$$\alpha \text{ at } T \,°C = \alpha \text{ at } 20 \,°C \times [1 - 0.000471(T \,°C - 20 \,°C)]$$

where 0.000471 is a constant for sugar-type samples.

Some useful terminology:

- **Specific rotation**, or specific rotatory power, in units are rad m²/kg; note that 2πrad = 360° (i.e. degrees of angle).

$$[\alpha]20_D = \frac{1,000a}{lc}$$

where α is the observed angular rotation degrees; ℓ is the length (pol) of tube in decimetres, usually 100 mm (i.e. 1 decimetre) or 200 mm; c is the concentration of compound in g/100 mL or kg/m³ (i.e. density); D denotes the sodium D line (at 589 nm) as the light source; 20 denotes the temperature of the measurement at 20 °C.

- **Ventzke** scale (Schmidt Haensch): 26.05 g, i.e. 1 V = 0.3466° angular rotation.
- **International Sugar Scale** (ISS): 26.000 g (at sodium light of 589 nm) i.e. 1 S = 0.34620° angular rotation.

  The **ISS** (°Z) is where the 100-point scale is defined by rotation of exactly 26.0000 g pure dry sucrose dissolved to a volume of 100.0 mL pure water (i.e. the **normal** sugar solution) at 20 °C, at wavelength 546.2 nm, with tube length of 200.0 mm; to yield 100 S where pure water is 0 S.

  At wavelength of yellow sodium light of 589 nm, the 100 S will be equal to the rotation of 34.6°.

  The ISS is linearly divided, i.e. a rotation of +17.31° (13 g/100 mL) equals to a reading of 50.00°Z.

  The 0°Z point in ISS is fixed by the indication given by the saccharimeter for pure water.

- International Commission for Uniform Methods of Sugar Analysis (ICUMSA) scale, i.e. the ISS scale in °Z
  i.e. 100°S = 99.971°Z
  or °Z = 0.99971°S
  thus 1°Z = 0.34626° angular rotation
- **French Sugar** Scale: 16.269 g, i.e. 1°FSS = 0.21667° angular rotation.
- **%Purity** can be read on a saccharimeter (a general-purpose polarimeter used in the sugar industry).
- **Schmitz's table:**

$$\text{Polarisation} = \frac{\text{Normal Wt.} \times \text{pol reading}}{\text{Wt. 100ml Soln in air}} = \frac{\text{Normal Wt.} \times \text{pol reading}}{100 \times \text{Apparent Density @ 20°C}}$$

The interrelationship between angular degrees and the °Z scales are defined by a linear equation.

For the determination of sucrose in the presence of other sugars, or vice versa, use the **double** polarisation method of **Clerget–Hertzfeld**. Examples of rotation other types sugars and substances, is shown in Tab. 10.3 below.

Tab. 10.3: Specific rotations of some substances, $[\alpha]_D$ in $°cm^3/g/dm$.

Optically active substance (aqueous solution)	Specific rotation at 20 °C
Dextrose (d-glucose)	+ 52.7°
Maltose	+137.5°
Fructose	−93.8°
Invert sugar	0°
Sucrose	+66.5°
Lactose	+55.3°
Ascorbic acid	+21.0°
Camphor (in alcohol)	+44.3°

There are generally three measuring ranges of polarimeters (saccharimeters):
–   +90° to −90° angular degrees
–   +180° to −180° angular degrees
–   +355° to −355° angular degrees (or −225 to +225 ISS or °Z).

Calibration of the instrument is done with quartz control plates or quartz wedges of known (standardised) degrees of optical rotation, at a specified temperature and light source.

Chapter 11
Turbidimetry and nephelometry

Keywords: turbidity, nephelometry, Tyndall effect, detectors, turbidimeters, units, errors in turbidimetry

This chapter outlines the various ways of measuring degrees of haziness, hue, turbidity and matter in suspension by passing light beam through the item or sample (liquid) at different angles.

The basic **principle** here is that of measurement of light scattering of a particle, which depends very much on the size of the particle and the angles of the scattered light to the incident light beam. The stability and temperature of the media in which the particles are suspended is critical; hence, measurement conditions must be rigidly adhered to in order to permit valid comparisons between samples and standards.

The measurement of the turbidity (haziness) of a liquid sample (usually water and wastewaters) is compared to a standard reference solution, such as a formazin 4,000 nephelometric turbidity units (NTU) suspension. This is useful for quality control work as well as measuring the stability of emulsions over a time period; example would be cosmetic liquids or gels.

In analytical chemistry, another common routine test is measuring trace levels of sulphates in liquids by creating a suspension of barium sulphate from barium chloride:

$$BaCl_2 + Na_2SO_4 = BaSO_4 + NaCl$$

Turbidity in a liquid is caused by the presence of finely divided suspended particles or colloidal matter in the liquid causing the **scattering of light**. The matter stays in suspension due to various electric charges. The intensity of this light (from an incident light beam) passed through the turbid sample is reduced by scattering (reflection of the light of these particles in all directions), and the quantity of light scattered is dependent upon the concentration and size distribution of the particles.

For calibration of the measuring instruments, various chemical standard formulations are used.

McFarland equivalence turbidity standards are used as standards in adjusting densities of bacterial suspensions.

NTU values are used as calibration standards (usually suspensions of formazin polymer) for nephelometers and turbidity meters.

Jackson turbidity units are roughly equal to NTU units.

Which units are used is dependent upon the method of measurement.

In **nephelometry,** the intensity of the scattered light (reflected) is measured (usually at **right angles** to the incident light beam), while in **turbidimetry,** the

https://doi.org/10.1515/9783110721201-012

intensity of light transmitted (not absorbance) through the sample is measured. Nephelometric measurements at 90 degrees incident light (**Tyndall effect**) are suitable for the low range (below 1,000 NTU), whilst turbidimetric measurements at 180° incident light are for higher ranges of NTU.

Both principles, nephelometry and turbidity, of measuring the intensity of light scattering are generally stated in NTU units. Various combinations of reagents are used as suspensions for calibrating these instruments. Generally, these calibration standards are not stable for lengthy periods of time (from several weeks to 2 years depending upon the formulation and manufacturer).

The incident **light source** of benchtop turbidimeters is usually a tungsten lamp (incandescent), LED or laser and the **detectors** are a type of photoelectric cell having a spectral response in the visible region of 400–860 nm. The type of sensing cell depending upon the manufacturer of the instrument and could be of laser or fibre-optic design.

The turbidimeter measurement **range** depends upon the light source; this range can vary from 0 NTU (tungsten filament lamp) up to 10,000 NTU (LED infrared source).

Depending upon the manufacturer, the turbidity measurement from a tungsten light source is measured at wavelength 450–600 nm whilst that from LED is measured to 860 nm. The USEPA 180.1 states the use of the tungsten lamp whilst the ISO 7027 states the use of the LED light source.

If instrumentation is not available, then a series of **formazin polymer standards** can be prepared in the laboratory and the sample visually compared against these standards.

Common errors in turbidimetric work, other than the instability of the calibrating solutions and the prepared test sample suspension, are the presence of entrapped air bubbles and presence of colour.

Another problem is where the sample containing cells, cuvettes or bottles is scratched or scuffed. This problem can be somewhat solved by a very thin coating of silicon oil which is rubbed over the scuffed markings to even out the glass surface of the container; excess oil to be removed.

There are many papers on the Internet where UV–vis spectrophotometers and colorimeters have been used to measure the turbidity of water or other liquids. A relationship exists between chromaticity coordinates and turbidity.

Chapter 12
Polarography (voltammetry)

Keywords: oxidation states, specificity, amperometric titration

The principle of polarography is the measurement of the **diffusion-controlled current** flowing in an **electrolysis cell** in which one electrode is polarised. The current is directly proportional to the concentration of an electroactive species.

Polarography is one of the few techniques that can be used to differentiate between the different oxidation states of the element (analyte); a tin example is stannous (Sn^{2+}) and stannic (Sn^{4+}); another example is the common one with iron, namely ferrous (Fe^{2+}) and ferric (Fe^{3+}).

Polarography is not ideal for qualitative analysis as the half-wave potential of the elements is not strictly characteristic for a specific element. However, for quantitative work, it can be measured in ppm levels.

Amperometric titration (**dropping mercury electrode**):

The dropping mercury electrode is a working electrode made of mercury and is used in polarography. Because mercury is a metal, but a liquid at normal temperatures, it can be renewed after each droplet. Thus, the working electrode is often a drop suspended from the end of a capillary tube.

The procedures are similar to those corresponding to voltammetry work which uses solid electrodes.

https://doi.org/10.1515/9783110721201-013

Chapter 13
Refractometry

Keywords: Wave velocity, refractive index, refraction, Snell's law, Abbe number, anisotropic

The outline of this chapter is the study of the **principles** of measurements of the bending of light, factors affecting the determination of the refractive index (RI) and its various uses in industry and laboratories.

The RI or index of refraction of a substance or **medium** is a measure of the **speed (wave velocity) of light** beam through that substance or medium (dependent upon the density of that substance or medium). It is expressed as a ratio (of angles) of the speed of light (VA) in medium A (generally air, but by definition vacuum), relative to that in a medium B (VB) (the sample being measured).

For example, the RI of **water** is 1.3330 at light of wavelength 589.3 nm, meaning that light travels 1.333 times as fast in vacuum as it does in water.

Some other examples are ethanol 1.361, glycerol 1.473 and 50% sugar 1.420. Note that these values are indices and hence have no units.

In forensics, the RI of a glass shard found at a crime scene can identify the source or type of glass; e.g. crown glass 1.517, flint glass 1.655. Also, it is very useful in gemmology.

The RI of a drop of essential oil can indicate its identity and quality.

The RI of a drop of cane or beet syrup can indicate its concentration of soluble sugars, aka Brix reading.

Refractometry can also be used to measure **film thickness** of materials by measuring the interference fringes. This is done using differential refractometry.

Where a is angle of incident light (relative to right angle to surface of medium B and b is angle of refracted light (Fig. 13.1).

The relationship between the angle of incidence and the angle of refraction, is expressed by **Snell's law of refraction**:

$$n = \sin ØA/\sin ØB$$

where n is the RI of medium A (usually air).

At a certain angle of incidence ØA, one part of the light beam is reflected and the other is refracted exactly along the interface of the two media. This angle is known as the critical angle of refection.

RI of substances varies with the wavelength of light. This is called dispersion and results in a slightly different RI for each colour (wavelength) of the incident light.

https://doi.org/10.1515/9783110721201-014

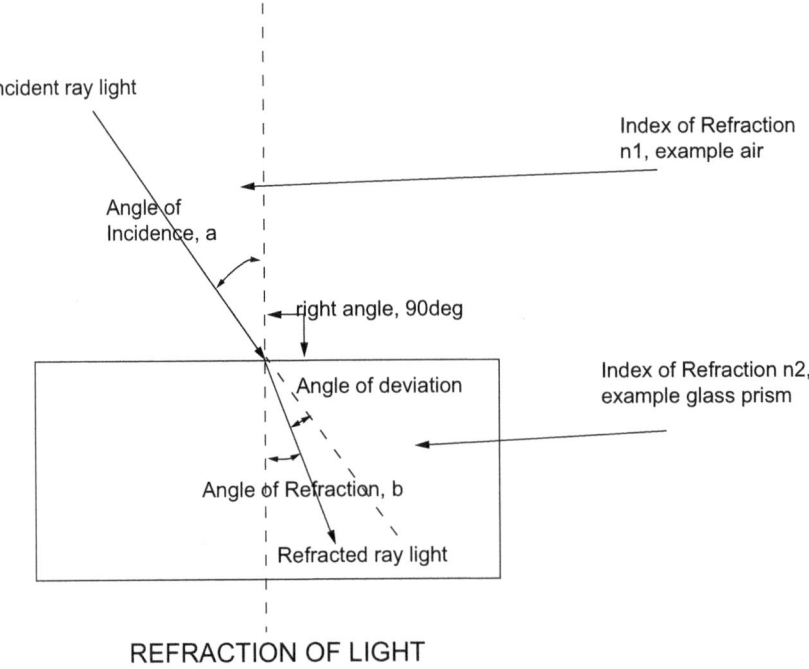

REFRACTION OF LIGHT

Fig. 13.1: Refraction of light rays.

RI or **Abbe number** can be measured at different wavelengths ranging from 450 to 1,550 nm.

As shown below:

– RI 1.3278 to 1.7379 (wavelength 450 nm);
– RI 1.3000 to 1.7100 (wavelength 589.3 nm), the standard wavelength of sodium D-light;
– RI 1.2912 to 1.7011 (wavelength 680 nm);
– RI 1.2743 to 1.6840 (wavelength 1,100 nm).

Abbe number, also known as the **V-number** or constringency of a **transparent** material, is a measure of a medium's **dispersion** (variation of RI with wavelength) in relation to the **RI.**

Abbe numbers are used to **classify glass** and other optically transparent materials. For example, **flint glass** has V of less than 50 and **crown glass** has V of greater than 50.

RI of substances varies with temperature. An example is the RI of essential oils where the temperature correction factor is generally taken as 0.00035 $°C^{-1}$.

Anisotropic solids (crystalline solids) are those that have more than one RI besides other physical properties. Sometimes, these are man-made, but generally they are various types of minerals and ores.

The most common type or design of refractometer is the Abbe refractometer. There are also available automatic refractometers, sometimes referred to as Brix refractometers. For field use, in agriculture, use is made of portable or pocket-type refractometers.

The RI of a transparent solid, such as diamond, can be easily determined by placing it in a series of liquids of known RI and viewing the solid in each liquid, until the solid becomes invisible in one of the liquids (at that stage, the RI of the solid will equal the RI of the liquid).

Chapter 14
Spectrophotometry

Keywords: electromagnetic spectrum, wavelength, frequency, radiation, units, absorbance, extinction, transmission, Lambert–Beer's law, atomic spectroscopy, colorimetry, colour scales, AAS, ICP, reflectance, fluorescence, UV–vis, detectors, calibration, UV and visible solvents, cut-off wavelengths, polar solubility, non-polar solubility, infrared, FTIR, ATR, IR absorbance peaks, flame types, gases, AAS sensitivity, flame emission spectra, flame absorption spectra, ICP limits

The outline of this chapter is the discussion of principles involved and the study of the many ways that light or energy measurements at different wavelengths can be used for identification and measurement of concentration levels of analytes and compounds.

The **principle** of spectroscopy is observing, producing and analysing or evaluating the spectra formed by substances when they are exposed to electromagnetic radiation from various sources.

Examples of instrumental emission spectroscopy techniques for determining elements in a sample are arc spark, dc arc, flame photometry, X-ray florescence, atomic emission spectroscopy, neutron activation analysis and mass spectrometry.

The electromagnetic spectrum (ems) or radiation is the range of wavelengths (length from peak to peak), frequencies (cycles per second) and amplitudes (electron volts), as explained in Fig. 14.1 below.

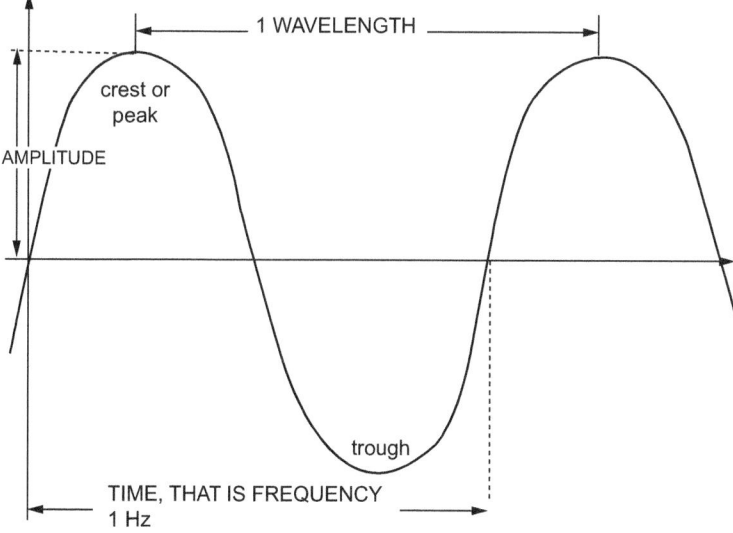

Fig. 14.1: Wavelength.

https://doi.org/10.1515/9783110721201-015

The **principle of spectroscopy** is the measurement of the interaction of the atoms, ions and molecules, functional groups of a substance with applied or emitted electromagnetic radiation. Examples are shown in Tab. 14.1 below.

Tab. 14.1: Electromagnetic spectrum.

Type radiation	Wavelength (m)	Frequency (kHz) approx.	Molecular and atomic activity
Gamma rays (short wavelengths)	1×10^{-12} to 10^{-14}	1×10^{18} (high frequency)	Unstable nuclear isotopes
X-rays	1×10^{-11} to 10^{-10}	1×10^{17}	Inner shell electrons
Ultraviolet rays	1×10^{-8} (190–380 nm)	1×10^{15} to 10^{16}	Outer shell (valency) electrons
Visible light: Violet Indigo Blue Green Yellow Orange Red	1×10^{-6} (400–420 nm) (420–440 nm) (440–490 nm) (490–570 nm) (570–585 nm) (585–620 nm) (620–780 nm)	1×10^{14}	Outer shell (valency) electrons
Infrared rays	1×10^{-6} to 10^{-4}	1×10^{12} to 10^{14}	Molecular vibrations
Microwave	1×10^{-3}	1×10^{10} to 10^{12}	Molecular rotations and spin
Radar	1×10^{-2}	1×10^{9} to 10^{10}	Nuclear spin
FM radio	1×10^{1}	1×10^{8} to 10^{9}	–
TV signals	1×10^{1}	1×10^{7} to 10^{8}	–
Shortwave radio	1×10^{2}	1×10^{6} to 10^{7}	–
AM radio (long wavelengths)	1×10^{4}	1×10^{5} to 10^{6} (low frequency)	–

Radiation comes in two forms, namely electromagnetic radiation (such as light) that has a range of frequencies from radio waves to gamma rays and travels at the speed of light (3×10^8 m/s) which is a stream of massless particles known as photons. The other form is known as the wave-particle theory based on kinetic energy levels. This form is usually depicted similar to an alternating current (AC) (of electrons i.e. electricity), having frequencies (Hertz, amplitude) and wave lengths (energy levels). Basically, shorter wavelengths yield higher frequencies (higher energy levels).

The **energy** of electromagnetic radiation is **inversely proportional** to the **wavelength** and hence the **frequency** of the wave is inversely proportional to the wavelength.

Units and scale of radiation used in spectrophotometry:

Wavelength:

- 1 Angstrom $= 1 \times 10^{-7}$ cm $= 0.1$ nm
- 1 m $= 100$ cm $= 1,000$ mm $= 1,000,000,000$ nm

Wavenumber is the inverse of wavelength measured in cm^{-1}.

Frequency is cycles per second, generally noted in kHz or MHz.

Ionisation potential is in eV.

Spectrophotometry is the quantitative measurement (-metry) of light (-photo) spectra properties of substances of their reflection, transmission, emission or absorption patterns, as a function of wavelength, wave number or frequency.

Terminology:

- **Absorptivity** is a measure of how much of the radiation is absorbed by a substance.
- **Reflectivity** is a measure of how much of the radiation is reflected by a substance.
- **Transmissivity** is a measure of how much of the radiation passes through the substance.
- **Emissivity** is a measure of how much of the radiation energy can be emitted by a substance to its environment.

The basic **principle** of spectrophotometry is the measurement of the amount of the absorption by a substance and the radiant energy (e.g. light) of different wavelengths (e.g. 560 nm) being applied to that substance.

Some common terms used in spectrophotometric work:

- **Absorbance** $= A = \log_{10}(I_0/I) = \log 1/T = abc$
  where I_0 is incident light; I is transmittance or transmitted light;
  a, b and c are constants;
  a is **absorptivity** of sample (it is a characteristic value for each substance at a stated wavelength);
  b is cell thickness in cm;
  c is concentration in g/100 g = density;
  Note: **Absorbance** is a logarithmic scale.
- **Extinction** $= E = \log_{10}(I_0/I) = A$.
- **%T** = Transmission (aka **transmittancy**) = antilog (2 − absorbance);
  or A = log 100/%T.
  Note: **%Transmission** is an arithmetic scale.
- **Transmission** = T = ratio is the radiant power transmitted by the sample to the radiant power incident on the sample;

- **Transparency** = I_t/I_i
- **Absorptivity** = a = A/bc,
  where b is the length of light path, i.e. size cuvette; and c is the concentration in grams per litre of the substance.
- **Molar absorptivity** = A where (c = mol/L; b = in cm).
- **Optical density** is same as absorbance A, which is a logarithmic scale.
- **Specific absorbance** = A where (c = 1%; b = 1 cm).

Lambert–Beer's law:

$$A = \varepsilon \times L \times C$$

where A is absorbance (has no units); ε is molar absorptivity (aka extinction coefficient), in units of L/mol/cm;

L is length of light path (generally in centimetres); C is concentration of the absorbing species of the analyte (chromophore), generally in grams per litre (density).

The law basically states that the **quantity of light absorbed** by a substance (i.e. a substance completely dissolved in a solvent, which is a solution without any turbidity present) will be **directly proportional to the concentration** of the substance (solute) or inversely proportional to the logarithm of the transmitted light.

This law deviates from its definition when the standards' calibration curve becomes non-linear; it could be a positive of negative curvature due to chemical interferences, physical interactions or stray radiation; especially when the absorbance value becomes too high such as greater than 0.5 absorbance.

The law is applicable to inductively coupled spectrophotometry (ICP), atomic absorption spectrophotometry (AAS), infrared (IR), near-infrared spectrophotometry (NIR), ultraviolet and visible spectrophotometry (UV–vis).

Typical **working ranges** for **atomic spectroscopy** instrument techniques are:
- Atomic absorption **flame** (AAF) is from 200 ppm down to 0.5 ppm.
- **ICP-OES – radial** is from 100 ppm down to 0.1 ppm.
- **ICP-OES – axial** is from 10 ppm down to 0.01 ppm (the axial spatial arrangement of the torch usually always has a lower detection limit than radial.
- **Hydride generation** AAS is from 0.1 ppm down to 0.005 ppm.
- **Graphite furnace** GFAA, is from 1 ppm down to 0.005 ppm.
- **ICP-MS** is from 1 ppm down to below 0.001 ppb.

Note that depending upon the manufacturer's instrument and how well the analyst has optimised the instrument's variables, the ranges above could be considered to be approximate linear working ranges.

Note:

1 ppb = 1 µ/L = 0.001 ppm = 0.0001 g/100 g or 0.0001% m/m

1 part per billion = 1 µg/L = 0.001 ppm = 0.0001 parts per 100.

14.1 Colorimetry

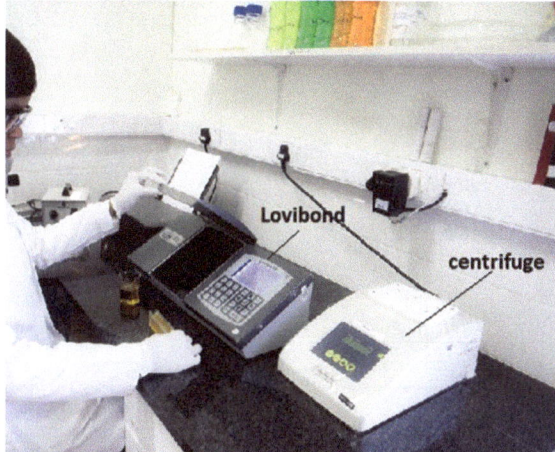

Fig. 14.2: Lovibond spectrophotometer.

The principle is basically the measurement of the visible light energy before and after being transmitted through a medium (liquid). A typical instrument is the Lovibond Spectrophotometer which measures intensity of colours of solution in various color scales.

The technique is sometimes referred to as photometry.

When white light is directed through a prism, it diffracts into the seven colours of the rainbow; this is the simple form of spectrophotometry. Today, diffraction gratings and other patented structures are used to split the light into a spectrum.

To remember the colours of the rainbow, use the easy acronym read over your green book in verse; red orange yellow green blue indigo violet.

White light passing through a prism is refracted (dispersed) into its component (rainbow) colours; this is the **basics** of simple spectrophotometry. This is due to the principle that the light beam is split into its parts (wavelengths, as shown in Tab. 14.2), by the fact that each light component (band) will travel at different speeds through the glass medium; because glass has a higher density than air and each wavelength has different energy levels.

Tab. 14.2: Colours of specific wavelengths.

Wavelength range in nm (approximate)	Visible colour
190–420	Ultraviolet region
400–465	Violet
465–480	Blue
480–500	Blue-green
500–530	Green
530–570	Yellow green
570–580	Yellow
580–590	Yellow-orange
590–600	Orange
600–620	Reddish-orange
620–780	Red
780–900	Near-infrared region

14.2 Colour scales

- **CIE L*,a*,b*** (aka CIELAB or LAB) is a scale where L* indicates perceptual lightness and has values between 0 and100; colours which have no chroma have coordinates a* = b* = 0; where a* is the redness or greenish with values between 0 and +80 and where b* is the yellowness or blueness with −80 to 0 values. Thus, International Commission on Illumination (CIE) values cover the range of RGBY (red, green, blue and yellow).
- **Lovibond** colour, RYBN: the instrument was initially based on 84 coloured glass filters, used for vegetable oils and fats, polymers, paints and so on; Lovibond also has portable comparators that use a disc wheel of standard colour filters.
- **Hunter Lab** colour: similar to the CIE scale but the algorithms used are different.
- **Klett colour** (the Klett–Summerson instrument is very rarely used): the scale is sometimes used for measuring bacterial flora suspensions.
- **Saybolt** colour: ASTM D156 and ASTM D 6045 are used in petroleum and mineral oil colour measurements, scale from −16 (dark colour) to +30 (light colour).
- **ASTM colour:** D1500 is used in petroleum and mineral oil colour measurements.
- **APHA** or **Pt–Co** (platinum–cobalt scale) aka **Hazen** units: ASTM D1209 is used in water industry; scale from 0 ppm (pure water white) to 500 ppm (yellowish colour liquid).
- **Gardner scale:** Used for vegetable and cooking oils.

- **Rosin** (ASTM D509) is used for solid semi-transparent materials, colour from yellow to reddish orange, where XC is brightest to D is darkest.
- **EBC** is the European Brewery Convention, where the light is measured at 430 nm wavelength; similar is the ASBC scale.
- **ICUMSA** colour refers to the sugar and syrup industry;
- **Brunswick colour** refers to the colour of sugar crystals ranging from mill white to brown.
- **Pfund equivalents** is the colour scale for measuring commercial honey.
- **FAC colour** is used for colour grading of fats (animal), tallows (animal grease) and oils.
- **ADMI colour index** refers to the ASTM 2120E standard; this index is basically used to measure the colour of effluents (wastewater from dye factories).
- **IP 17:** used for petroleum products and oils.

Note that:

ASTM stands for the **American Society for Testing Materials**

IP stands for the **Institute of Petroleum**, UK

APHA stands for the **American Public Health Association**

ICUMSA stands for the International Commission for the Uniform Methods of Sugar Analysis.

FAC is the colour scale used by the American Oil Chemists' Society for fats and oils

ADMI stands for the American Dye Manufacturers' Institute

There are many websites on the Internet that offer online **conversions** from one colour scale to another, but these are just **approximations** as only the bespoke instrument can measure the required colour of a liquid accurately.

14.3 Ultraviolet and visible spectrophotometry

The **principle** of this technique is the measurement of light energy in the UV and visible range of the ems, namely from 190 nm to about 350 nm (UV range) and from 300 to 1,100 nm (visible range). The range depends upon the manufacturer, model of the instrument and its internal optics and software. The isolation of a wavelength for a determination depends upon the variable or fixed, slit width or spectral bandwidth of the instrument; this could range from 20 nm (low-cost instruments) down to research-type instruments of 0.5 nm spectral tolerance (resolution) of the selected measuring wavelength.

The effect of slit width (bandwidth) is crucial when interpreting absorption spectra that are to be used in qualitative determinations as the sought narrow wavelength of the unknown substance could be lost or hidden, if the narrowest slit width is not selected.

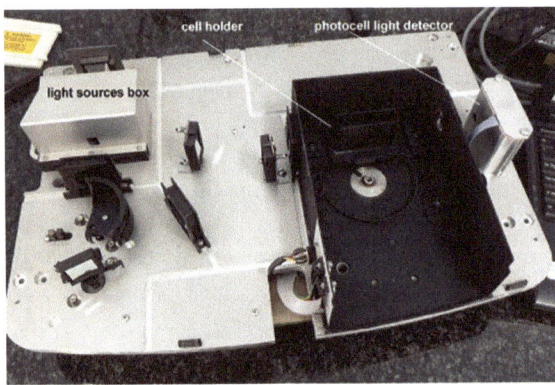

Fig. 14.3: Inside a simple UV–vis spectrometer.

Most of these spectrometers on the market today can undertake a variety of tasks, such as full wavelength scanning to produce a UV–vis spectrum, continuous data measurements during reaction or enzymatic changes at various temperatures, kinetic studies. Temperatures can be controlled by the use of Peltier units.

Accessories are usually automatic cell changers, a range of different cuvette holders, optical fibre probes for out-of-compartment **remote measurements** and **reflectance** measurement probes.

The light or **energy sources** (a typical layout of a spectrometer is shown in Fig. 14.3 above) are ordinary candescent bulbs (tungsten), or light emitting diode for the visible range and a hydrogen lamp (deuterium) for the UV range. Some instruments make use of a xenon flash lamp.

The **measuring detectors** are photomultiplier tubes, photodiode arrays and many similar photosensitive units depending upon the manufacturer and various patent design rights.

The **analytical uses** of this type of instrumentation are very similar to that of basic colorimetry or photometry, but of a wider detection wavelength range and has a lower more sensitivity and selectivity of colour hues. The width of the bands (wavelength range) is commonly controlled by the narrow band pass of the instrument.

Normally, samples are liquid because the incident light energy source is directed through the liquid which then absorbs some of the light radiation, which is then measured as energy absorbed or transmitted. The energy scale can be in units of absorbance (A) or %transmittance (T).

Disadvantage of this technique is that it has low or no specificity; but is highly sensitive to particular chromophores of different molecular structures and is simple to use.

Some spectrophotometers have a **double beam operation** which basically is where the incident beam of light from the light source is split and transmitted through the sample cell and simultaneously transmitted through a reference cell. The reference cell can contain a standard solution or a blank reagent or background corrections.

Other types of instruments are where there are two beams of light, of which one beam is transmitted through the sample cell and the other beam through the reference cell. Other types might have two beams and, in addition, two detectors.

The type of measurement of a double beam spectrophotometer will produce more stable, reliable and precise readings.

Many spectrophotometers come with an adjustable **band pass** (as mentioned above) or slit width, usually 5 nm for the less costly instrumentation and about 1 nm for more sophisticated and research type instrumentation. The narrower the band pass, the more precise the nearest 1/10 of a wavelength of the spectrum can be measured and broad peaks do not hide smaller peaks when wavelengths are at very close proximity to each other.

Some spectrophotometers are capable of scanning in emission mode and the UV–vis range. This is a useful technique when establishing what the major unknown components are present in the sample solution being atomised.

To verify or **calibrate the accuracy** (or misalignment of the wavelengths, stray light or spectral resolution) of a spectrophotometer, a series of dilutions of a solution of **potassium dichromate** may be used in place of external reference filters, such as holmium oxide or didymium glass.

Potassium dichromate in sulphuric acid yields a spectral scan of characteristic peaks at 235, 257, 313 and 350 nm. These are usually measured at an instrument's bandwidth (slit width) of 2 nm.

Prepare a solution of 0.060 g potassium dichromate, dissolved in 0.005 M sulphuric acid and make volume up to 1,000 mL. Measure the absorbance in a 1 cm cell. Readings should be as below:

235.0 nm = 124.5 A
257.0 nm = 144.0 A
313.0 nm = 48.6 A
350.0 nm = 106.6 A

The common **holmium oxide glass** reference filter has a number of sharp absorption bands which occur at precisely known wavelengths in the ems from 250 to 700 nm. This reference filter is usually built into the instrument for purposes of **self-calibration**.

Another way of describing Lambert–Beer law, aka Beer–Lambert law, relates to the **attenuation** of light to the properties of the medium through which the light is transmitted.

$$A = \alpha bc$$

where A is absorbance; α is absorptivity, which depends on factors such as cell path length, solvent, temperature and slit width; and c is concentration.

This attenuation or absorption of light is dependent upon the concentration of the medium. This is generally a linear function at low concentrations, but at higher concentrations the relationship becomes nonlinear.

The **linearity of this law** is limited by chemical and instrumental factors such as electrostatic interactions between molecules at close proximity to each other (i.e. at higher concentrations, the medium becomes denser) and changes in the refractive index (RI) of the medium at higher density of the medium.

Spectrophotometric solvents (refer to Tab. 14.3 below) to be used in the instruments' measurements must comply with the following:
- They must be almost completely transparent at the wavelength of interest;
- The use of certified reference material or "Spectro" grades should be used at all times to avoid the misconception of the presence of impurities or unknown peaks.

Tab. 14.3: Solvents for UV–vis spectrophotometry.

Solvent	Polarity	Dielectric constant (ε)	UV cut-off (lower wavelength limit) in increasing nm; in a 10 mm cuvette
Water	Polar protic	78	185
Acetonitrile	Polar protic	37	190
n-Pentane	Non-polar	2	190
Hexane (UV grade)	Non-polar	2	195
Glycerol	Polar protic	47	200
Cyclopentane	Non-polar	2	200
Cyclohexane	Non-polar	2	205
Methyl alcohol	Polar protic	33	205
Isopropyl alcohol	Polar protic	18	208
Ethanol	Polar protic	24	210
2,2,4-Trimethylpentane (iso-octane)	Non-polar	2	210
Tetrahydrofuran (THF)	Polar aprotic	8	212
Ethyl ether	Non-polar	4	215
n-Butyl alcohol	Non-polar	18	215
1,4-Dioxane	Non-polar	2	215
Ethanol			220
Dichloromethane (DCM)	Non-polar	9	233
Chloroform	Non-polar	5	245
Ethyl acetate	Polar aprotic	6	260
Methyl formate	Polar aprotic	9	260

Tab. 14.3 (continued)

Solvent	Polarity	Dielectric constant (ε)	UV cut-off (lower wavelength limit) in increasing nm; in a 10 mm cuvette
Toluene	Non-polar	2	265
Carbon tetrachloride TOXIC	Non-polar	2	265
Dimethyl formamide (DMF)	Polar aprotic	37	270
Acetic acid	Polar protic	6	270
Dimethyl sulfoxide (DMSO)	Polar aprotic	47	270
m-Xylene	Non-polar	3	280
Benzene (fraction) HAZARDOUS	Non-polar	2	285
Dioxane			320
Carbon disulphide EXTREMELY FLAMMABLE	Non-polar	3	330
Pyridine HAZARDOUS	Polar protic	12	330
Acetone	Polar aprotic	21	330

The **UV cut-off wavelength** is where the solvent itself will absorb all the light below that wavelength. Note that generally, the UV range extends from 100 to 400 nm, depending upon the manufacturer's construction parameters, such as optics and software.

The **basic principle of polar and non-polar solubility is like dissolves like!**

Non-polar (dielectric constants less than about 5) solvents have bonds between atoms that have related electronegativities, whilst **polar** (they have dielectric constants greater than 5) solvents have atoms with very different electronegativities between them (such as hydrogen H^+ and oxygen O_2^-).

Thus, the dielectric constant of a solvent is a measure of its polarity. The higher the value, the more polar is the solvent.

The **dielectric constant** is actually a measure of the amount of electric potential energy (e.g. compare capacitors as to how they store electricity), in the form of induced polarisation that is stored in a given volume of substance under the action of an electric field.

Polar solvents can be either protic (i.e. capable of hydrogen bonding, –OH) or aprotic (that is, it has no –OH or –NH groups).

When doing UV spectrophotometry measurements, ensure that the cuvettes (aka cells) are quartz and not glass or plastic (acrylic). Note that plain glass or plastic absorbs UV radiation.

Note that acetone solvent attacks plastic ware.

14.4 Infrared spectrophotometry

The **principle** of IR spectrophotometry is that it measures molecular energies or frequencies of vibration of organic materials (molecules), whereas AAS or flame emission spectrophotometry (FES) spectrophotometry measures atomic absorbance or emission energies of inorganic or mineral materials.

The types of samples that IR can handle are powders, thin films, solids and liquids.

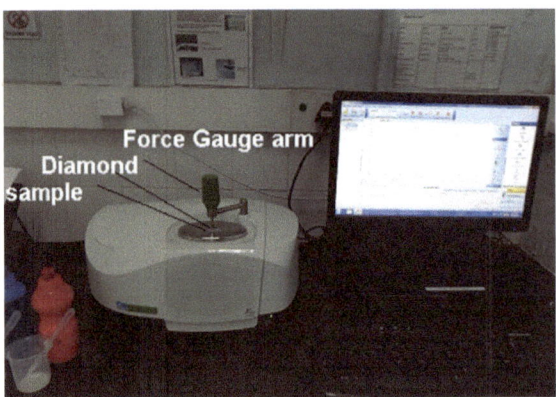

Fig. 14.4: FTIR spectrophotometer.

14.5 FTIR spectrophotometry

Fourier transform IR (FTIR) transformation is the principle where the signal to noise ratio is mathematically enhanced by mean of filters or an array of splitters and mirrors. The outcome of this FT modification is the speed of scanning of all frequencies (wavenumbers) simultaneously from one end of the IR spectrum to the other.

IR spectroscopy is generally used in qualitative analysis to separate, identify organic groupings or molecular bonding structures. The instrument will display spectrum of absorbance (Tab.14.4), emission or %transmission scale for solids, liquids and gases.

To convert from %transmission to absorbance:

$$A = 2 - (\log_{10} T)$$

IR absorbance spectroscopy obeys the Lambert–Beer law whereby band absorbance is linearly related to low concentrations of the molecules in solution.

IR spectrophotometers use the polystyrene IR spectra to calibrate and verify the instrument. It is an ideal substance to easily observe the wavenumber (wavelength, aka frequency) of the peaks and the peaks' intensity.

Accessories to a FTIR spectrophotometer is the **attenuated total reflectance (ATR)** where a sample of solid or liquid is placed (clamped or held down) on a

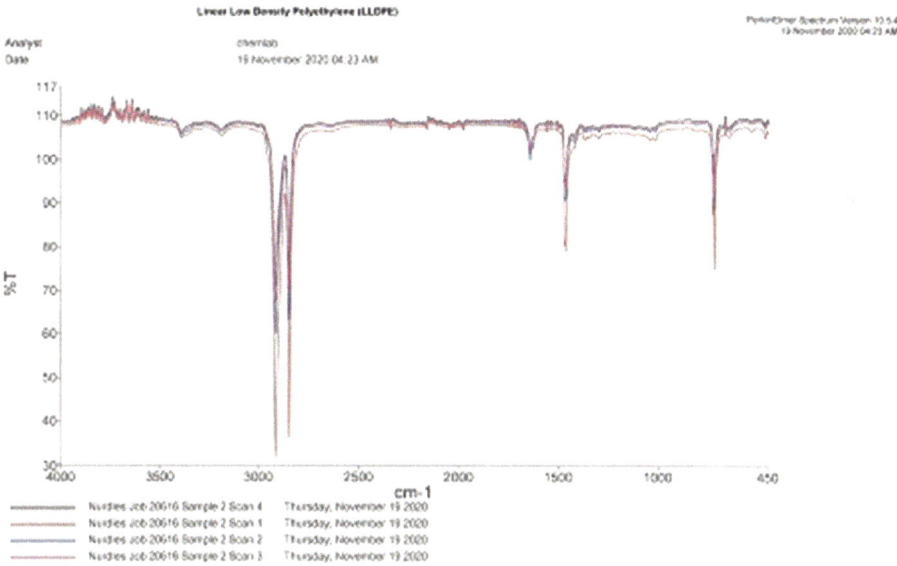

Linear Low Density Polyethylene (LLDPE)

Fig. 14.5: FTIR spectrum.

crystal and its IR spectrum is measured, without any need to prepare KBr or NaCl salt discs. This is the **major advantage** of using the ATR principle that no longer requires the tedious sample preparations of the traditional IR spectroscopy by **transmission** of the light beam. An example of this type instrument is shown in Fig. 14.4 above. A typical spectrum from this instrument is shown in Fig. 14.5 above.

The ATR accessory (refer to fig. 14.6 for a diagram of the principle the ATR) measures the changes that occur in a totally internally **reflected** IR beam when it comes into contact with a sample. The beam is directed onto a **diamond** type material (optically dense crystal of a high RI which must be greater than that of the sample's RI) at a **certain angle**. This internal reflectance creates an evanescent wave that extends beyond the surface of the diamond into about 5 microns depth of the sample, which is held in direct contact (avoiding any entrapped air) with the crystal surface.

In the regions of the IR spectrum where the sample absorbs radiant energy, the **evanescent wave** will be altered (attenuated). This alteration or attenuation energy is passed back to the beam which then exits by the opposite end of the crystal and is hence passed to the detector of the spectrometer.

An evanescent wave is where energy is spatially concentrated in the vicinity of the source.

The crystal or diamond must be kept clean and free of any dust particles. It can be cleaned with a soft tissue damped with acetone, or methanol or isopropanol.

IR spectroscopy can be and is used in quantitative analysis to a certain degree of accuracy.

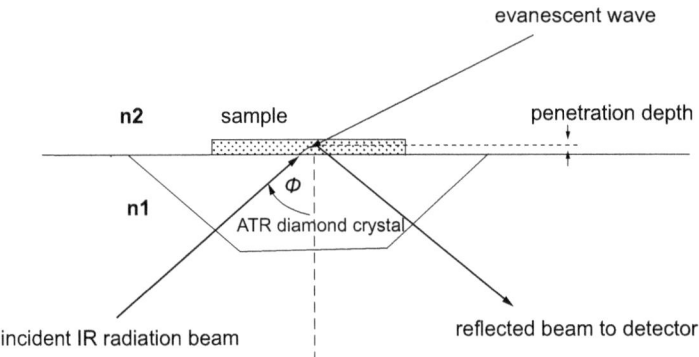

Fig. 14.6: ATR accessory.

The quantitative analysis can be done by comparing the peak height or peak area of the absorbed lines compared to (or superimposed onto) a standard spectrogram.

The general **advantage** of FTIR over common IR spectrometers is that it can measure the intensity of the radiation over a wider spectrum range.

Disadvantage is that water should not be used as solvent or be present in the sample to be read on the instrument. However, this can be overcome by running a background scan on pure water.

The absorbed radiation is converted into vibrational stretching (symmetric and asymmetric), vibrational bending (twisting, rocking, wagging, scissoring) and rotational energy of the sample's carbon bonding functional groupings. The wavenumber (inverse of wavelength) at which this absorbance (or Transmission) occurs, will identify the type of bonding and grouping of the sample under investigation. This spectrogram can then be compared to a published library of spectrograms in an attempt to identify an unknown sample makeup or detect impurities.

For example, methane (CH_4) has absorptions at $3,000$ cm^{-1} and $1,310$ cm^{-1}.

Mid-IR spectroscopy is an established technique for fingerprinting of various solids, liquids or gases that are used in various industries, such as paint and forensics. Substances can be characterised, identified and quantified.

The general range of the most commonly used spectrophotometers are:
– Far-IR: 400 cm^{-1} to 30 cm^{-1}
– Mid-IR: $4,000$ cm^{-1} (2.5 µm) to 400 cm^{-1} (25 µm)
– Near-IR: $14,000$ cm^{-1} to $4,000$ cm^{-1}
– FTIR: $8,000$ cm^{-1} (1.25 µm) to 350 cm^{-1} (29 µm); depending upon manufacturer and model
– Fingerprint region: $1,400$–600 cm^{-1};
– Functional group frequency region: $4,000$–$1,300$ cm^{-1}.

The light source is generally a tungsten halogen lamp whose energy is then bombarded at the sample's molecules, which are then absorbed.

The **detectors** used, depending upon the manufacturer, are photoelectric or pyroelectric sensors.

Tab. 14.4: IR absorbance of major peaks.

Functional group	Wavenumber band, in cm^{-1}
C–H	+–3,000
C=O	1,630–1,850
O–H	3,400–3,650
Carboxylic acids O–H	2,500–3,000
Aromatics (benzene ring)	1,600–2,000
C≡N	2,100–2,260
C≡C	2,100–2,260
C–O	1,050–1,150
C=C	1,620–1,680
Alcohols, phenol: O–H	3,200–3,550
Amines N–H	3,500–3,700
Nitriles C≡N	2,220–2,260
Aldehydes and ketones C=O	1,740–1,690
Carboxylic acids C=O	1,710–1,780
Amides N–H	3,500–3,700
Amides C=O	1,630–1,690

ATR IR is the latest type of instrumentation, where the advantage is little, or no sample preparation is required. **ATR** spectra are similar but not identical to the normal IR spectra.

14.6 Near-infrared spectrophotometry

NIR is a spectroscopic method that principally measures the light absorption of a substance, from about 780 to 2,500 nm of the ems, when subjected to a light source (commonly halogens or LED).

The **advantage** of this instrument over that of an IR instrument is that it does not require special IR radiation as a light source of energy. The other advantage of this technique is that the measurement is non-destructive and very little sample preparation is required.

The **disadvantage** is that it is not a suitable qualitative tool, due to the few absorption peaks.

The other (major) disadvantage is that many samples (>50) of like kind have to be tested by referee methods, in order to calibrate the instrument.

Most common use is in the food and agriculture industry for measuring moisture, proteins, fats and carbohydrates. Other fewer common uses are measuring levels of alcohols and carboxylic acids.

14.7 Raman spectrophotometry

The principle of Raman spectrophotometry is the identification of the symmetry of molecular functional groups in organic compounds by using light scattering of the excited vibrations.

Lasers are usually used as the light excitation sources.

The spectre is useful for the identification of the structure and symmetry of molecular groupings.

The technique is applied in the structure determination and identity of organic compounds and the symmetry of molecular groups in solid state.

The advantage in qualitative analysis is the identification of vibrational modes of functional groups, usually in water solutions, that are usually different from those identified by IR spectroscopy.

The technique can also be used to provide structural fingerprint by which molecules can be identified.

The basic **difference** between IR spectrophotometry and Raman spectrophotometry is that the former measures the excitation of molecular vibrations by light absorption, whereas the later measures the excitation of molecular vibrations by light scattering.

Disadvantages are that sample must not fluoresce and solution must not be turbid nor coloured.

14.8 Mass spectrophotometry

Mass spectrometry measures the **mass-to-charge ratio** of molecules using **electric** and **magnetic** fields. In other words, it detects and measure the abundance of the molecular fragments derived from the sample under examination.

There are several ionisation methods and mass analysers, depending upon the manufacturer's design and the specific end use of the instrument.

Mass spectrophotometry (MS) is basically, an instrument that can be used on its own for ionisation (i.e. breaking up) of groups of organic molecules or functional grouping into further sub divisions (fragments).

MS can also be used as a detector attached to the output signal from another instrument, such as a GC; hence, the measuring principle would be GC-MS, thus assisting in the identification of unknown compounds or molecular structures.

The technique is useful for the molecular weight determinations of the fragments of the molecules of organic compounds.

MS is not suitable for very stable non-volatile samples.

14.9 Atomic absorption spectrophotometry

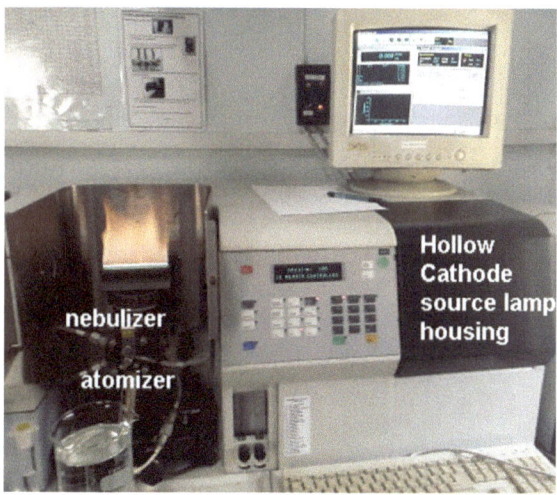

Fig. 14.7: AA air–acetylene flame.

The principle of flame spectre was based on the simple (flame emission) instrument, namely a flame photometer, where the unknown substance is burned in a flame and the colour emitted (589 nm wavelength) identifies the chemical element, e.g. the appearance of bright yellow flame when a pinch of salt on a spatula is placed over a flame such as a Bunsen burner. This principle is commonly used as a quick and simple forensic check for presence of various materials.

The **flame photometer** uses a cold flame, such as butane gas and the sample in liquid form, is sprayed (atomised) into the flame; the wavelengths of the light emitted is split by means of filters and is measured by a galvanometer or photocell. This instrument is commonly used for detecting and measuring alkali and alkaline earth elements. The atomiser or nebuliser is of a type known as a total consumption burner head.

The simple flame photometer uses an ordinary incandescent or LED as a light source; whereas in AAS, the most common light source is the **hollow cathode lamp (HCL)** where the cathode is made from the same element as that sought for the analyte. These lamps can be multi-element types.

The **principle** of atomic absorption is where an atom (free atom) of interest (the analyte) is bombarded with energy rays from a light source (HCL) of the same wavelength as the atom itself. The atom absorbs this energy by using it to move its electrons from its ground state to the next higher energy level. The amount of energy absorbed is measured and is proportional to the level of concentration of the free atoms of that analyte.

Basically, what this means is that any element in the periodic table can absorb or emit light under certain specified conditions of energy transfer. The amount of light absorbed or emitted is proportional (linear or nonlinear) to the concentration of the analyte atoms having their valence electrons being excited (ionised in ICP spectrometers).

A basic **AAS instrument** (refer to Fig. 14.7 for example of a typical AA instrument in operation) consists of:
− Light source (HCL) of the same wavelength (radiant energy) as the element to be determined (analyte)
− Atomiser to create a mist of fine particles of the liquid (solution) being sprayed into the flame (burner head)
− Nebuliser to create an aerosol of a nuclear mist within the high temperature combustion gases
− Monochromator to separate the characteristic energy wavelength from all the other wavelengths being emitted by the light source, this is usually by an arrangement of optics (lens and mirrors) consisting of a grating as well as a slit width gate
− Detector to measure the absorption response from the light passing through the nuclear active part of the flame
− For solid-type samples, an attached electrothermal (graphite) furnace is used in place of a flame
− For heat-sensitive elements such as mercury and arsenic, a hydride generating accessory is used to create a cold stable vapour before it enters the AAS light path

Other associated techniques are flow-injection units or attachments, digital display of data, calibration graphs and results, with attached printers or databases.

AAS is very effective to detect and measure trace amounts of metals in lubricating and transformer oils, extracted in organic solvents such as methyl isobutyl ketone (MIBK) and methyl ethyl ketone. Organic solvents can, under certain conditions, enhance (increase flame temperature) the signal.

Also applicable in flame photometry (FES) measurements.

Generally, the **air–acetylene flame** (about 2,200 °C) is used, but for refractory metals (e.g. Al and V) where a hotter flame is required (higher energy), the **nitrous oxide/acetylene flame** (about 3,000 °C) is required. For a list of the flame types and their temperatures, refer to Tab. 14.5 below. Other more sensitive techniques are hydride generation as is used for arsenic and selenium, whereas the **cold vapour technique** is used for mercury compounds.

In practice, there are two types of premix linear burner heads used, the 50 mm slot (definitely essential for stability and safety of the nitrous oxide/acetylene flame) and the 100 mm slot.

Some AAS use the total consumption burner, but this type is more suitable for the common flame photometer.

To increase sensitivity, that is lower the detection limit, use the 100 mm burner head to decrease sensitivity use the 50 mm burner head (using this, burner will cut light path length in half, hence decrease the dilution factor, thus magnitude of error). Other examples of decreasing the sensitivity, i.e. to reduce any dilution error or to measure samples higher in concentration of the analyte, are:
– burner head can be turned at an angle so as to reduce the light path length through the flame (this is only possible on some models of AA (atomic absorption) instruments; or
– use a less sensitive wavelength, e.g. for Ca use 239.9 nm in instead of 422.7 nm; or
– reduce the HCL energy (current); this will also extend the life of the HCL.

Some AA spectrophotometers have the facility of automatically scanning which is useful when doing qualitative work by flame emission photometry (a flame emission spectrum).

Most AA instruments can measure not only in the absorbance energy mode, but also in the atomic emission energy mode. This can be an advantage as the alkali and alkaline earth metals exhibit better sensitivity.

The AAS technique is sensitive for most of the metals in the periodic table; however, it is prone to chemical, spectral and sample matrix interferences. Other problems are the incomplete atomisation of the analyte atoms, or inversely the unwanted ionisation of the analyte atoms. Various ways have been developed to mitigate these errors or enhance sensitivity and maximise absorbance, such as:
– using the general analytical technique of standard additions;
– hydride atomisation;
– addition of chemicals, such as lanthanum nitrate or lanthanum chloride; strontium can be used in place of lanthanum;
– atomising at higher temperatures, e.g. removal of calcium interference in the air–acetylene flame by using the hotter nitrous oxide flame.
– fine tuning of the pneumatic nebuliser to mitigate the incomplete or insufficient atomisation;
– using a cooler flame to mitigate the ionisation of the analyte atoms;
– measuring the analyte at a different spectral wavelength. Refer to Tab. 14.6 for emission and Tab. 14. 7 for absorption spectra, below.
– narrowing the bandpass of the instrument if possible, depending upon model and make of the spectrophotometer.

- running the HCL lamp at a higher current; this will reduce its life;
- aligning the HCL lamp manually to obtain highest directed energy output to the photodetector;
- ensuring that the burner head is not blocking the light beam from the HCL. This can be assisted by placing an old business card on top of the burner head in the centre (obviously with the flame off!) to block the light beam, then align the burner head vertically (rotational) and horizontally until the light spot is central on the card; this technique can also be used to align the HCL light beam.
- note that optimising flame profile (fuel: oxidant) gas ratios, even burner height; should be optimised manually to obtain maximum sensitivity with minimum background noise. Example is chromium which requires a fuel rich flame.
- flame AAS are used as the atomisers (formation of a mist) in atomic spectroscopy. The temperatures of the flames are important as they discern which elements would be capable of atomisation followed by levels of excitation of the valence electrons of the analyte atoms.

Tab. 14.5: Flame types and temperatures.

Flame atomisation	Oxidant	Fuel	Temperature, °C (approximate)
Household gas	Air	Butane pentane blend	1,800
Air–hydrogen	Air	Hydrogen	2,000
Air–acetylene	Air	Acetylene	2,300
Nitrous oxide–acetylene	Nitrous oxide	Acetylene	2,700
Oxy–acetylene	Oxygen	Acetylene	3,000

Note that any acetylene gas leaks can be detected by the characteristic odour of carbide.

Note that any nitrous oxide gas leaks can cause drowsiness, as it is commonly used as (anaesthetic) laughing gas.

Tab. 14.6: Flame emission spectra.

Element	Wavelength lines in nm
Ag	328, 338
Ba	554, 744, 873
B	250, 345, 452, 548
Bi	223
Ca	423, 554, 622

Tab. 14.6 (continued)

Element	Wavelength lines in nm
Co	**346**, 353, 387
Cr	361, **425**, 428
Cs	456, **852**, 894
Cu	325, **327**, 520
Fe	**373**, 386
K	345, 405, **766**
Li	323, 460, **671**
Mg	**285**, 371, 383
Mn	280, **403**, 543
Na	330, **589**, 818
Ni	**342**, 353, 386
Pb	261, 368, **406**
Rb	420, **780**, 795
Sr	408, **461**, 821

Tab. 14.7: Flame atomic absorption spectra.

Element	(less alternative sensitive wavelengths in brackets)	
	Absorption wavelength line in nm	Sensitivity (concentration in µg/mL to yield 1% absorption)
Al	309.3 (396.1)	0.8???
Ag	328.1	0.03
As	193.7	0.8
Au	242.8	0.1
B	249.8	8
Ba	553.6 (350.1)	0.2
Be	234.9	0.02
Bi	223.1	0.2
Ca	422.7 (239.9)	0.02
Cd	228.8 (326.1)	0.01

Tab. 14.7 (continued)

	(less alternative sensitive wavelengths in brackets)	
Element	Absorption wavelength line in nm	Sensitivity (concentration in µg/mL to yield 1% absorption)
Co	240.7	0.07
Cr	357.9 (425.4)	0.06
Cs	852.1	0.1
Cu	324.8 (327.4)	0.04
Fe	248.3	0.06
Ge	265.2	2
Hg	253.7	2
In	303.9	0.4
K	766.5	0.01
Li	670.8	0.02
Mg	285.2 (202.5)	0.003
Mn	279.5 (403.1)	0.02
Mo	313.3	0.3
Na	589.0	0.003
Ni	232.0 (341.5)	0.07
Pb	217.0 (283.3)	0.1
Pt	266.0	1
Sb	217.6	0.3
Se	196.0	0.5
Si	251.6	2
Sn	224.6	0.5
Sr	460.7	0.04
Te	214.3	0.3
Ti	364.3	2
V	318.5	0.9
W	255.1	6
Zn	213.9 (307.6)	0.009
Zr	360.1	9

Note that AAS can handle aqueous and some organic solvents, such as MIBK. Organic solvents are used to determine metals in used oils and petroleum products.

Some analytes have to be complexed with sodium borohydride in order to lower their detection limits.

Atomic absorption techniques can be summarised as below:

- **Electrothermal atomisation (GFAA)** is where the graphite furnace is used to detect elements at the nanogram levels (ppb). It can be used for solids and liquids which are vapourised and atomised.
- **Hydride generation(AAS)** atomisation techniques using sodium borohydride are used generally for microgram levels (ppm) of volatile metals such as antimony, selenium, arsenic and tin.
- **Cold vapour atomisation (AAS)** technique is used to detect and measure nanogram (ppt) levels of mercury.
- AAS instruments which have the facility for **FES measurements** can be used for quick and easy qualitative investigations if the instrument model has a scanning capability and the software incorporates that facility in its programme or coding algorithm. These emission spectrograms are useful in identifying unknown specimens.
- **Nitrous oxide** acetylene flame atomisation;
- **Air acetylene** flame atomisation.

14.10 Inductively coupled spectrophotometry

ICP does not replace AAS but is used to complement that technique as some analytes are more sensitive to AAF than the plasma gas torch of the ICP. The limits of the various techniques is shown in Tab. 14.8 below.

Sodium is more sensitive and stable when measured by atomic emission than by atomic absorption.

ICP has proven to be useful for refractory metals (elements) such as Si, Ti, B, W and Mo.

It can also detect total oxides of sulphur and phosphorus compounds. It is useful also for halogens.

The instruments basically consist of:
- Computer software
- Optical system
- RF power supply
- Nebuliser
- Spray chamber
- Auto-sampling device
- Cooler unit supply to the torch housing
- Supply of gas, e.g. argon

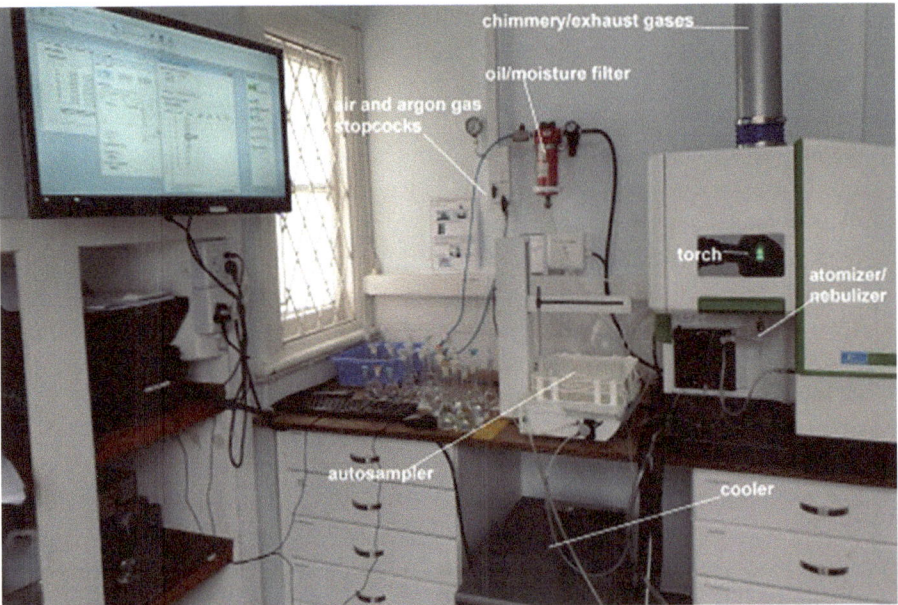

Fig. 14.8: ICP spectrophotometer.

Tab. 14.8: ICP limits.

Instrument type	Dynamic range (µg/L)	Detection limit (µg/L)
GFAA	Max 3	0.001
Flame AAS	Max 4	1
ICP-OES radial	Max 7	0.1
ICP_OES axial	Max 6	0.01
ICP_OES dual	Max 10	0.01

The **temperature** of the argon gas plasma (ionised gas) is about double that of the AA air–acetylene flame.

The **argon gas** is bombarded with a high radio frequency energy source, hence making it flammable as a high-temperature torch (flame). The energy (temperature) of the torch excites the analyte's element (atom) of interest to its next high atomic level.

ICP-OES (where OES is optical emission spectrophotometry) (a typical ICP-OES instrument setup is shown in Fig.14.8 above) has poor sensitivity levels for antimony, arsenic, mercury and selenium compounds, thus these should ideally be isolated as the hydrides and then measured by ICP-OES. A typical ICP instrument settings is shown in Fig. 14.9 below.

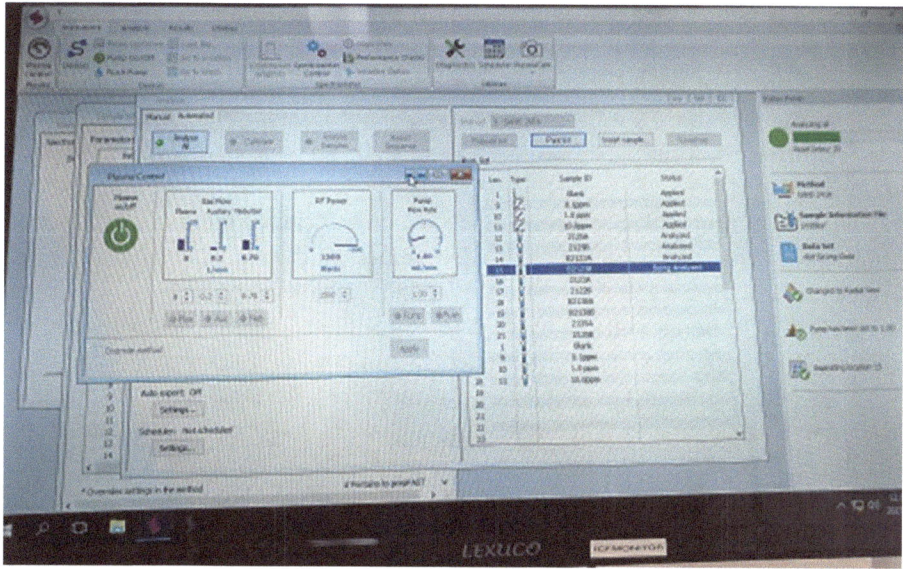

Fig. 14.9: ICP software screen.

Most ICP instruments offer two types of positions of the light beam through the plasma torch, namely, axial or radial.

Axial optical viewing offers greater **sensitivity** (as much as ten times more than radial) as the direction of the (incident) light beam is parallel to the plasma (from above the torch), thus offering a larger amount of light signal to the detector. **Disadvantage** is more background noise and matrix interferences (e.g. from sodium and potassium) leading to poor precision.

Radial optical viewing is where the light beam is **perpendicular** to the longitudinal axis of the plasma torch; hence its path through the plasma is shorter, offering less sensitivity but better precision, of the analyte's metal element. The radial type does offer a wider dynamic **linear** measurement range.

Radial is also more amenable to **non-aqueous** solutions (organic solutions).

Another more costly instrument is the **dual**-type viewing, which combines both views in one instrument, hence incorporates the advantages of axial and radial types.

The maximum dissolved solids content of solutions should not exceed 25% (**radial** type can handle higher dissolved solids up to 30% TDS).

14.11 Reflectance spectrophotometry

Reflectance spectroscopy or photometry is the observation or measurement, of the spectral composition of the **surface-reflected** radiation with respect to its angularly dependent **intensity** and the composition of the incident primary radiation from that surface of the substance under investigation.

Since most substances absorb IR radiation, it becomes sometimes necessary to measure the substance's reflectance characteristics instead. However, reflectance spectra (other than the ATR) can appear very different from their transmission spectra.

When radiation strikes a surface of a substance, it can be reflected, transmitted or absorbed. The relative amounts are determined by the refractive indices of the two media (usually air and the substance) and the angle of incidence.

Diffuse reflection spectra in the **mid-IR range** (2–20 μm) are generally used for qualitative identification (molecular-fingerprint) of powdered coatings.

There are instruments available that will measure reflectance and transmittance spectra in a single process. The understanding here is that of light scattering (diffusion) spectroscopy.

The spectrophotometric range is from about 250 nm to about 850 nm, against a white standard of 100% reflectance to a black standard of zero reflection or maximum light absorption.

This type of measurement of colour intensity and reflectance is used in fields such as biology, medical science, food technology and biochemistry.

There are many variants of industrial colour measuring instrumentation available, based on slightly different configurations.

14.12 Fluorescence spectroscopy

This technique, also known as **fluorimetry**, has basically two advantages over UV–vis spectrophotometry, namely specificity and sensitivity. Specificity because the principle of fluorometry is that it measures both the absorption and fluorescence at an optimal wavelength of an analyte, and lower sensitivity.

Fluorescence spectroscopy measures the fluorescence (a type of luminescence emitted from a substance when a beam of UV light excites the electrons in its ground state to its outermost shell of that substance or compound. Not all substances are able to emit the fluorescent light in the UV–vis wavelength range, which can be detect and hence measured.

This technique is useful in the art world where detection of illegal copies of paintings can be detected.

Fluorescence and phosphorescence are emission processes where atoms or molecules are excited by absorption of a beam of electromagnetic radiation and produce radiant energy which is measured when the excited species return to their original or ground state. Phosphorescence takes a longer time to return to its ground state.

14.12.1 Physical measurements

This section outlines the most common physical test procedures that are undertaken in most analytical laboratories, such as viscosity, density, particle size grading and surface tension.

Other common tests considered for physical testing or for properties of a sample (item) are flash point, fire point, upper explosive limit and lower explosive limit, auto-ignition temperature, melting or pour point, cloud point, freezing point, vapour density, boiling point and distillation range.

Knowledge of the physical properties of materials and chemicals is required in order to design efficiently and safely, build, produce, transport or store, these substances. Hence, this becomes part of an analytical chemist's tasks as physics, chemical engineering and chemistry are closely intertwined.

The chemist also has to compile and produce **safety data sheets** for every chemical product that is manufactured and transported throughout the world. This is a mandatory requirement of many organisations, such as the IMDG code and IATA regulations.

The analytical chemist may also be called upon to do other physical measurements (in addition to those mentioned above) which are necessary when it comes to compiling various technical and chemical engineering documents such as designs, patents, drawing plans and upscaling from laboratory-scale process to a factory development.

These could include:
- Vapour pressure
- Freezing point depression (cryoscopic measurement)
- Boiling point (ebulliometric measurement to minimise superheating whilst boiling solutions in a flask)
- Boiling point elevation of a liquid by adding another liquid to increase the boiling point of the initial liquid (or solvent)
- Melting point (measuring the purity of crystals or their identification)
- Identifying the seven different geometric forms of a crystalline substance: triclinic, monoclinic, orthorhombic, tetragonal, trigonal, hexagonal and cubic
- Surface tension (measuring effectiveness of emulsifiers and detergents)
- Magnetic properties (e.g. identifying materials, minerals or ores)

Physical measurements

Chapter 15
Viscosity

Keywords: units, kinematic, dynamic, Newtonian, non-Newtonian, rheometry, pseudoplastic, thixotropic, capillary viscometers, viscometers, viscosity index

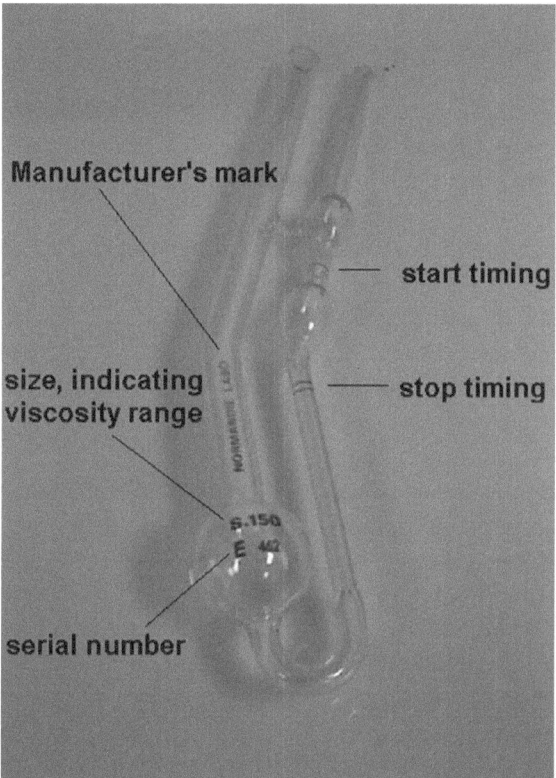

Fig. 15.1: Cannon–Fenske routine glass viscometer.

This chapter outlines the various types of apparatus or instrumentation that is used for measurements of the resistance to flow or change of shape or form of fluids and the instruments used.

The **principle** of viscosity is the measurement of the resistance to flow or movement of a fluid under certain conditions of stress. In other words, viscosity is the measurement of internal friction or resistance, to flow or movement, by external forces.

Viscosity is thus the ratio of shear force to shear rate:

https://doi.org/10.1515/9783110721201-016

$$N = \frac{f/A}{dv/dx} = \frac{\text{shear force}}{\text{shear rate}}$$

where N is the absolute viscosity, f is force, A is area, dv is change in velocity and dx is change in distance

Viscosity is generally reported in **centipoises** or **centistokes,** sometimes in seconds of flow rate:

Centipoises = Centistokes × Density of the liquid.

1 centipoise = 1 millipascal second

Some other units are **Saybolt Universal** (SSU) and **Engler Seconds,** e.g. 100 SUU = 150 Engler Seconds (note the relationship is not linear).

The viscosity of a fluid is greatly affected by **temperature.**

Hence, all viscosity measurements must be undertaken at a specific temperature and this temperature must be controlled to within 0.05 of a degree or better.

Thus, a very accurate and calibrated thermometer must be used for temperature control whilst performing viscosity measurements. Also, the temperature must be constant throughout the measurement!

Accurate and consistent timing devices (time recordings to within 0.1 s) are also imperative when applicable to certain viscometer measurements.

There are commonly two types of viscosity, namely:

- **Kinematic viscosity** where the measurement is made while the fluid flows under the normal force of gravity; the viscosity units are usually cSt (centiStokes, cSt), square-metre per second. An example is the capillary-type viscometers (Fig. 15.1):

$$1\ cSt = 10^{-6}\ m^2/s$$

- **Dynamic viscosity** is viscosity of the fluid based on the fluid's density, and the viscosity units are usually cP (centiPoises, cP), or Newton-second per square metre. An example is the Brookfield viscometer:

$$1 cP = 10^{-3} N \cdot s/m^2$$

Note: cSt × SG = cP (both must be at same temperature for conversion of scale units).

There are basically two rheometry (viscosity) properties or categories of fluids: Newtonian fluids and non-Newtonian fluids.

Newtonian fluids (e.g. water and oil) are those whose stress (e.g. drag or resistance to flow) is directly proportional to the stress or strain being applied to it (e.g. force of gravity, drag or resistance to flow(friction)). They have the same viscosity at different shear rates, e.g. different rpm of spindle (Brookfield). Basically, Newtonian fluids are those whose viscosity is independent of any shear rate applied to it. Most materials or substances are of this type.

Non-Newtonian fluids (e.g. tomato sauce (ketchup), polymers and paints, corn starch) are those where its viscosity or pseudo-plasticity varies depending on the type of applied stresses. They have different viscosities at different shear rates.

There are two types: **pseudoplastic** (are independent of the time the stress is applied) and **thixotropic** (are dependent upon the time the stress is applied to it).

Thus, when reporting viscosity results for non-Newtonian samples, it is critical to state the time and manner of the applied stress; e.g. Brookfield spindle #6 at 30 rpm for 5 min.

Rheometry is the study of the flow and deformation of matter which describes the interrelationship between force, deformation and time. Rheometry includes the subject of viscosity.

Fluids have different rheological characteristics that can be measured by different types of viscometers. Rheometers are used for those materials which cannot be defined by a single viscosity value or measurement and hence require more physical parameters to be measured. It is commonly used to characterise the mechanical properties and the cross-linked structure of hydrogels and similar type materials.

There are various types of instruments that measure the viscosity of liquids:

kinematic viscometers (capillary flow under gravity) such as the glass capillary tubes of various dimensions, based on time taken for a specific volume of liquid to flow a specific distance under gravity.

A type where the time it takes an air bubble to rise to surface of a liquid, in a stand-ardised tube.

Another type is where the torque (resistance to movement) is the measure of a moving obstacle in a liquid.

Examples of kinematic **capillary glass tubes** are:
- Cannon–Fenske
- Ostwald
- U-tube
- Bs/U/M miniature U-Tube
- Reverse-flow type
- Ubbelohde
- SL suspended level
- Zeitsfuchs

and some less commonly used. These are mostly used in the petroleum industry. These have come in various sizes to accommodate varies sample volumes; the measurements are usually done at 40 and 100 °C temperatures.

From left to right of Fig. 15.2:
- U-tube viscometer
- U-tube reverse flow viscometer
- Suspended level Ubbelohde viscometer

– Cannon–Fenske viscometer; these can be reused for repeat measurements on same sample.

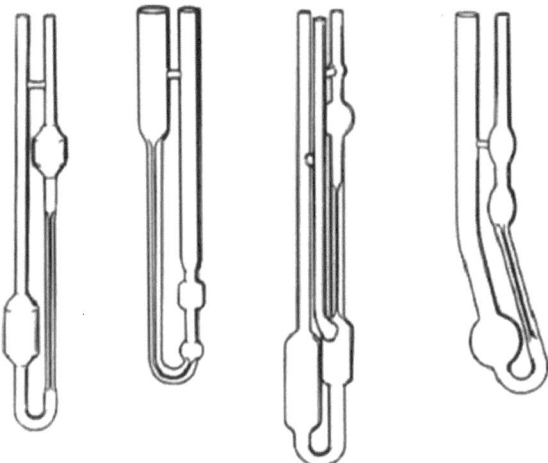

Fig. 15.2: Capillary glass viscometers.

Other types of viscometers are:
– **Hoeppler** viscometer (aka Haake **falling ball** viscometer) is based on the resistance of a falling body. It is measured on the time it takes a specific size and mass of a steel ball to fall a fixed distance through a fixed volume of liquid under gravity. Ideal for transparent Newtonian liquids such as inks, solvents, hydrocarbons and molasses.
– **Brookfield** viscometer and the **Haake** viscotester based on the drag of a rotating spindle are mostly used for emulsions and polymers; these rotating viscometers are used for both Newtonian and non-Newtonian-type solutions, gels or pastes.
– A similar type to the Brookfield is the **cone and plate** viscometer;
– **Benson** viscometers are a type of plasma viscometer usually used in pathology laboratories;
– **Ford cups**, these are similar to the now obsolete Redwood viscometers, where the time taken for a specific volume of liquid to flow through a specific aperture, under gravity, is measured.
– **Saybolt** Universal viscosity and Saybolt Furol viscosity are now obsolete, but many industries and engineers are still referring to its units;
– **Redwood** viscosity is now obsolete, but many industries and engineers are still referring to its units.
– **Gardner standard bubble viscosity** is where the time taken for a bubble to rise in a tube of liquid (sample).

There are published tables available for converting units of one type of viscometer to another type, but these are *only* approximate values, e.g. Brookfield to Ford cup.

Most types of viscometers must be regularly calibrated against certified reference fluids, because their characteristics (e.g. dimensions of the glass capillary type, mechanical parts of Brookfield) do change with continual use.

Viscosity index (VI) is an arbitrary measure of a fluid's (e.g. lubricating oils) change in viscosity relative to temperature change. VI has no units as it is an index value.

Published tables (ASTM DS 39B) are available for obtaining the VI for kinematic viscosity, in centistokes, measurements done in the laboratory at 40 and 100 °C, thus alleviating the necessity of calculating these values from the equations:

For oils of 0–100 VI:

$$U = L - \{(VI/100)\ (L - H)\}$$

For oils of 100 VI or greater:

$$U = antilog(\log H - N \log Y)$$

where Y is the kinematic viscosity of the oil at 100 °C; L is the kinematic viscosity at 40 °C of an oil of 0 VI and having the same kinematic viscosity at 100 °C as the oil whose VI is to be calculated; H is the kinematic viscosity at 40 °C of an oil of 100 VI and having the same kinematic viscosity at 100 °C as the oil whose VI is to be calculated; U is the kinematic viscosity at 40 °C of whose VI is to be calculated:

$$N = \log\{0.00715(VI - 100) + 1\}$$

The lower the VI of the fluid, the more the viscosity is affected by changes in its temperature.

Basically, the VI measures the ability of an engine oil, to resist becoming thinner at higher temperatures.

Chapter 16
Density

Keywords: density, SG, relative density, VCF factors, temperature factors, terminology, units, Archimedes' principle, pycnometer, density 10% solutions, hydrometers

Basically, the outline of this chapter is the measurement of mass and hence volume by determining the density of a substance, using specific type of bottles and hydrometers, and automatic instrumentation.

Density is, by definition, the mass per unit volume of a substance.

Naturally, density measurements are critically dependent upon temperature measurements because the volume of a substance will expand or contract depending upon the change in temperature of the substance.

The analytical chemist needs to know how to measure the density of various substances, solids, liquids and gases.

Density measurements are required in quality control procedures for determining the purity or concentration of substances. It can also be used to identify what the substance is by referring to the many various published density or SG tables on the Internet. Examples of densities of different type substances are shown in Tab. 16.1 below.

Density (**litre weight in air**) is used in the shipping industry to measure and control the loading of liquid cargoes (e.g. load line markings on the hull of a ship, known as the Plimsoll line, which, inter alia, measures the maximum cargo weight displaced by the density of the ocean water).

There are automatic density meters on the market, but these are programmed for specific types of substances depending upon their **temperature volume correction factors**.

Example of some VCF factors:
- Drying oils = 0.00068 per °C
- General fats and vegetable oils = 0.00064 per °C
- Linseed oil = 0.00062 per °C
- Gas oil (petrol) = 0.00067 per °C
- Fuel oil = 0.00063 per °C
- Jet fuel (paraffin) = 0.00072 per °C
- Ethanol (alcohol) = 0.00110 per °C
- Methanol = 0.00094 per °C

Note that the above factors will vary somewhat depending upon the purity of the substance.

Relative density is the same as specific gravity (SG) which basically is a ratio of two densities at two different or same temperatures; hence, it has no units.

https://doi.org/10.1515/9783110721201-017

The value of density is often related to that of water, where the density of water at 4.0 °C is taken to be exactly 1.0000 grams per cubic centimetre (g/cm^3).

The density of pure water at 20.00 °C is 0.998203 g/cm^3.

Apparent density is same as hydrometer density.

True density is same as density in vacuo.

Density in air (e.g. determined by use of the SG bottle or pycnometer) is same as litre weight in air.

The **units** of density measurements can be expressed in kg/m^3, g/mL, g/cm^3, g/cc, lbs/ft^3, g/L or hectolitre mass.

Density of gases can be used to determine molecular weight using the ideal gas laws:

$$Molecular\ weight = (density/pressure) \times gas\ constant \times absolute\ pressure$$

To determine the density of liquids in the laboratory, there are commonly three laboratory procedures, namely by hydrometer, by pycnometer or by automatic density meters. A new instrumental technique is that of using an oscillating u-tube.

A **pycnometer** is a vessel that can hold a known exact volume of fluid at a specific temperature.

To determine the **density of solids**, the use of **Archimedes' principle** applied to buoyancy and displacement is commonly applied. The principle states that when a body is immersed in a fluid, it will displace its own weight of that fluid.

The other alternative is to weigh the object, calculate its volume using mensuration formulae and then apply the definition of density, namely:

$$density = mass/volume.$$

Tab. 16.1: Density 10% m/m aqueous solutions.

Chemical solution	Density, true, at 20 °C (g/cm^3)
Ammonia (NH_4OH)	0.957
Ammonium chloride (NH_4Cl)	1.028
Calcium chloride ($CaCl_2$)	1.084
Copper sulphate ($CuSO_4$)	1.107
Ethanol (alcohol)	0.982
Formaldehyde (HCHO)	1.028
Glycerine ($C_3H_8O_3$)	1.022
Hydrochloric acid (HCl)	1.050
Methanol (CH_3OH)	0.982
Nitric acid (HNO_3)	1.054

Tab. 16.1 (continued)

Chemical solution	Density, true, at 20 °C (g/cm³)
Phosphoric acid (H_3PO_4)	1.053
Sodium carbonate (Na_2CO_3)	1.103
Sodium chloride (NaCl)	1.071
Sodium hydroxide (NaOH)	1.109
Sodium sulphate decahydrate ($NaSO_4 \cdot 10H_2O$)	1.040
Sulphuric acid (H_2SO_4)	1.066
Zinc chloride ($ZnCl_2$)	1.089
Zinc sulphate ($ZnSO_4 \cdot 7H_2O$)	1.058

16.1 Hydrometers

What is a *hydrometer*?

A **hydrometer** is a graduated glass or metal instrument used to measure either the specific gravitySG, API gravity, density or purity/strength of a liquid. It is based on the hydrostatic principle of the Greek mathematician and inventor Archimedes (Archimedes principle), which states that the mass loss of a body immersed in a liquid equals the mass of the liquid displaced. Most car battery hydrometers are enclosed in glass tubes fitted with rubber bulbs for drawing up the battery acid (into the tube) to be measured for its acid strength (sulphuric acid at SG of 1.275).

The hydrometer floats (buoyancy) in the liquid (it has a bulb-like bottom weighted with lead shot). Thus, when immersed in the liquid, the graduated stem rises vertically to show a scale reading. In order to take an accurate reading, the eye must be in level with the liquid surface (to avoid the error of parallax).

Most hydrometer scales are read as above, others are read at the top of the capillary surface. Usually, there is a coloured strip at the top of the stem or an image of the liquid attraction to the hydrometer stem (Fig. 16.1).

Hydrometers must be calibrated according to the **type of liquid** to be tested (because of **surface tension effects**) and at a standard temperature, usually 4 or 20 °C. There are specific types of hydrometers available for specific types of liquid products; refer to Tab. 16.2.

Factors to consider when using hydrometers to measure density (or levels of a solute (analyte) in a solvent (usually water) in a solution:

- Use of correct or applicable hydrometer to the solution or liquid being measured
- Temperature of the liquid

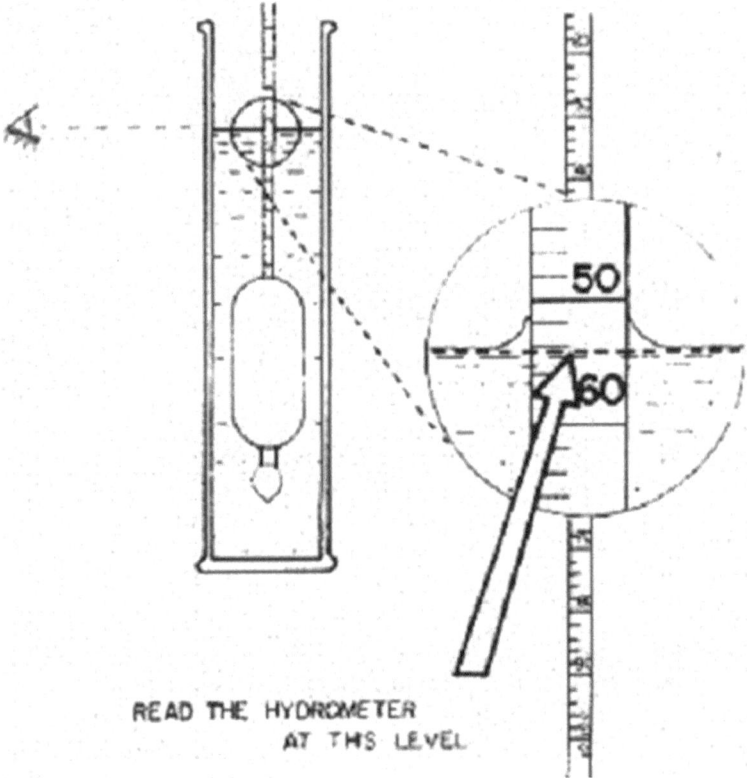

Fig. 16.1: How to read a hydrometer?

- Correct scale reading on the stem of the hydrometer; note that the scale reading is from lowest concentration (density) at the top of the stem, to the highest concentration lower down the scale
- Surface tension of the liquid
- Adsorption of the liquid to the surface of the material that the hydrometer is made; this can be prevented by giving the hydrometer a slight twist with the fingers when immersing it into the liquid being measured

The strength of fruit juices (Brix hydrometer) can also be measured by refractometer. Unfortunately, there is *no* exact correlation between these two instruments (hydrometers and refractometers).

Note the markings on the hydrometers, refer to Fig. 16.2 and 16.3 below:

M100 G/ML 20 °C BS 718 HIGH ST BT No.

- M100 g/mL indicates that its range is from 1.000 to 1.100 g/mL (density)
- BS 718 indicates that the hydrometer was manufactured according to British Standards number 718 (this document has been withdrawn)

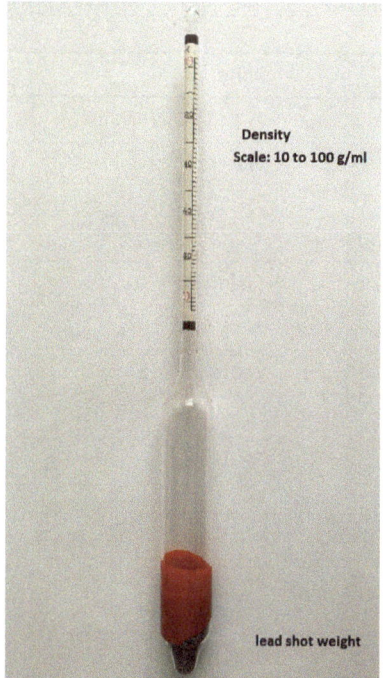

Fig. 16.2: A glass hydrometer with lead shot as a weight.

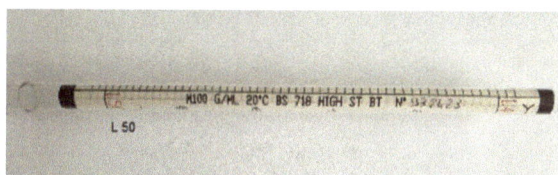

Fig. 16.3: Hydrometer scale.

- HIGH ST indicates that it is suitable for liquids of high surface tension (e.g. petroleum products)
- BT indicates
- No. is the serial number of the hydrometer; this indicates that this hydrometer has been calibrated.

The type of hydrometer to be used is dependent upon the physical properties (surface tension, density and turbidity) of the liquid being measured.

Examples:
- Low ST markings indicate measurements below 1.000
- Medium ST markings indicate measurements from 1.000 to 1.600
- High ST markings indicate measurements above 1.600

Tab. 16.2: Types of hydrometers.

Hydrometer	Some uses	Units
Alcoholmeter or Tralle	Alcohol content in beverages	% alcohol by volume
Syke's (or Cartier's), usually a brass type of spindle	Alcohol content	% alcohol by volume
Salinometer	Salt (NaCl) content in brine solutions	°S
Balling (Plato)	Sugar in brewing	Degree wort extract
Lactometer, Quevenne	Dissolved solids in milk or skim milk	%
Soxhlet's lactometer	Density of milk	25° to 35° (SG)
Brix or saccharometer	% sugar by mass in juices	°B
Twaddle	Densities greater than water, e.g. strength of H_2SO_4	°TW
Density	Density of solutions	g/mL or kg/L
Specific gravity (SG)	Relative density solutions greater than water	(no unit, as SGs are a ratio of two densities)
API Gravity	SG of petroleum products	°API
Baume'	SG of solutions	°Be'
Oleometer	Weight and density of vegetable oils	Degrees
Ammoniameter	Density ammonia solutions	0° to 40°
Urinometer	Urinalysis	°SG
Thermohydrometer	Hydrometer with built-in thermometer	°API

Note: °B is also Balling saccharometer units.

Density by definition is mass per unit volume, i.e. D = mass/volume. Hence, density values will always have units such as g/mL or g/cm³; also, density is measured at one temperature, whereas SG is expressed at two temperatures.

SG (aka relative density) is a **ratio of two densities** and hence its values have no units.

Note: Density and SG are *not* the same, except they have same numerical values *only*, where

$$\text{density} @ 20°C = SG @ 20°C/4°C$$

Because SG is a ratio of densities of two different substances, where in general, water is taken to be one of the substances; hence,

$$SG = \frac{\text{density of substance @ T\,°C}}{\text{density of water @ t\,°C}}$$

If temperature t °C of water is at 4.0 °C, then density of water = 1.00000000 g/mL. Therefore, the numerical values of SG @ T °C/4 °C

$$= \frac{\text{density of substance in g/mL @ T°C}}{1.00000000 \text{ g/mL}}$$
$$= \text{density substance in g/mL @ T °C.}$$

Thus, this is the only time when SG and density are of same value.
Other hydrometer conversions are:

$$SG = 1 + °TW/200, \text{ where TW is the Twaddle hydrometer;}$$
$$°TW = (SG-1) \times 200; \quad SG = 145/(145 - °Be)$$

$$°API \text{ gravity @ } 15.5\,°C = \frac{141.5}{SG\,@\,15.5\,°C/15.5\,°C} - 131.5$$

where API is American Petroleum Institute (for liquids lighter than water);

$$°Be = 145 - (145/SG)$$

where Be is the Baume hydrometer (for liquids heavier than water, e.g. brine);
°Be = 140–130 SG   (for liquids lighter than water, e.g. oils);
°Brix = % soluble solids = % sugar = grams of sugar per 100 grams of liquid.

Chapter 17
Particle analysis

Keywords: sieves, GPC, test sieve analysis, sieve conversions, mesh number, particle counters, imaging

The outline of this chapter is the study and measurement techniques used in the laboratory to obtain data on the shape, size and form of smallest particles such as particles of micron size, to large pieces of matter such as items of 5 mm size.

17.1 Gel permeation chromatography

Gel permeation chromatography is a type of molecular sieving chromatography, where a sample is separated into its constituent parts based on their molecular sizes, e.g. the characterisations of polymers by their molecular weight distributions.

Particle size distribution and analysis is undertaken in many industries, such as paint pigments and plastics, to check quality (quality assurance) of raw materials or of final consumer goods against standard international specifications, local trade legislation or factory manufacturing requirements.

There are various techniques for measuring the particle size distribution of materials. However, there is no single technique that can be recommended for all types of samples (Tab.17.1), since factors such as
– particle size,
– size range,
– particle shape,
– hygroscopic nature,
– chemical and physical stability of the particle and
– electrostatic probabilities and density have to be taken into consideration.

Example of some materials of different sizes:

Tab. 17.1: Size of common materials.

Material	Sieve mesh	Inches	Microns (µm)
Stones for concrete use	10	0.079	2,000
Beach sand	28	0.028	700
Fine sand	60	0.010	250

https://doi.org/10.1515/9783110721201-018

Tab. 17.1 (continued)

Material	Sieve mesh	Inches	Microns (μm)
Portland cement	200	0.003	74
Vegetative pollen	400	0.002	37
Red blood cell	1,200	0.001	12
Cigarette smoke	–	0.0001	2
Bacteria	–	–	1
Viruses	–	–	0.1

To measure particles smaller than test sieve 600 mesh (20 μm), use is made of automatic electronic counters, such as the Coulter counter, or dynamic imaging and light scattering instrumentation. The portable types are generally used for monitoring the environment in so-called clean room areas.

Other grading techniques are photographic optical granulometry.

For cell counting or that of vey minute particles, use is made of a light microscopy procedure using an hemocytometer slide with etched grid lines and calibrated against a linear scaled (micrometre) eyepiece.

17.2 Test sieve analysis

The principle of this most common particle size analysis is to separate (or even filter out) solid particles or powders into their specific particle size by using a series of screens or meshes of different shapes and various aperture sizes; each is for a particular purpose. Refer to Fig. 17.1 above as an example of stack sieves mounted on automatic shaker.

It is up to the experienced chemist to decide which type and size to use.

Sieves come in various materials, sizes and shapes, from the common round brass 200 mm diameter Tyler sieves to perforated metal plates.

Published tables are available for converting from one type, such as Tyler mesh number to BS 410, or ASTM E-11 to ISO 3310. There are no exact linear relationships between the different Standards or pore (aperture) nominal size.

Each standard or specification of a test sieve have their own particle aperture sizes and conversions;

Example:

0.500 mm pore size = 35 US ASTM E-11 = 32 Tyler mesh number = 30 BS 410

Fig. 17.1: Process of test sieving on automatic shaker.

Mesh number is generally referred to as number square apertures per inch. Other units are US mesh number and Tyler mesh number, refer to published tables for conversions; however, the two units are, for all real practical purposes, the same size opening.

Examples of types of sieves or screens, are:
– Woven wire mesh, 20 μm to 2.36 mm
– Square mesh wire, 500 mesh (0.025 mm) to 3.5 mesh (5.60 mm)
– Slotted plate openings; size of rectangle depends upon the industrial use, e.g. coal
– Perforated plate round hole or square openings, 3 mm and larger
– Dockage sieves for grain grading and similar materials

The standard frame sizes can be:
Circular of 3 in (100 mm), 8 in (200 mm) and 12 in (300 mm).

Frame materials are usually made of brass, bronze, stainless steel, aluminium or wood depending upon the required robustness of the sieving, shaking or vibrations of the screening proces.;

Square wooden frames are usually used for specific commodities such as for corn or wheat gradings.

Also, the separation can be done by **dry sieving** (free flowing powders) or **wet sieving** using water or non-corrosive liquids; depending upon the type of material particles being separated.

For example, food or organic materials would not be wet water washed through the screens as they would absorb the moisture, swell and then block the apertures of the screen.

Wet washing is often used to separate or sieve out mineral particles (e.g. aggregates such as stones, ores and sand) of different size. Another use of wet sieving is to wash out or clean water-insoluble materials for further lab analysis.

The size determinations or measurements are commonly referred to as **particle size distribution**, **screening**, **sieving** or **particle characterisation**.

The process involves shaking or vibrating a stack of screens (large test sieve sizes at the top of stack, then descending down to progressively smaller sieve sizes and finally the pan at the bottom of the stack) in order for the particles to pass through the apertures of the wire mesh screen. This can be done by hand sieving, or by automatic sieve shakers, timed generally at 5 min. The principle here is that the smaller particles pass through the pores of the sieve and the larger particles are retained by the sieve.

Over vibrating/shaking/ultrasonic vibrations on automatic sieve shakers can cause incorrect measurements as the particle grains start to grind against themselves, hence causing more smaller grain sizes than were originally present in the sample! This is a common error in routine factory laboratories.

To prevent this, manual sieving with gently brushing and tapping the sides of the sieve with the hands, and patience, should be undertaken by the laboratory analyst.

To obtain accurate and precise size distribution of materials (items or samples), the following needs to be taken into consideration:
- **Hygroscopic** properties of the material to avoid clumping or lumping of the particles
- **Friability** or robustness of the materials being evaluated or graded; here the time settings on automatic shakers are critical
- **Static** electricity can be produced during the shaking or vibration of certain materials, such as plastics and fine powders, causing the particles to adhere to the frame of the sieve or adhere together to form clumps
- Damage caused by the **abrasive** nature of the sample particles to the sieve construction materials

The difficulty is in determining the end or **finalisation** of the sieving process. This should be when there is no significant change in the mass of the test sieves in the stack. The word significant is important, because one can continue shaking or vibrating until doomsday, due to one or more of the above factors.

The test sieves must be inspected regularly as the solder seam of the mesh to the frame is often torn due to wear and tear or abuse in handling. The apertures or openings must also be clear of any previous materials or samples. The best method to clear the opening is gently brushing the mesh with a set of camel hairbrushes.

The data obtained from a stack of test sieves can be used to determine various properties of the material or sample, such as:

– Percentage retained on a sieve, i.e. the mass particles left on the mesh as a percentage of the whole sample mass.
– Percentage passed through the sieve, i.e. mass particles passed through the mesh as a percentage of the whole sample mass.
– Accumulated masses or percentage is the total or accumulated weight of all the weights of particles that have passed through the top layers of the stack. If the total percentage does not sum to 100, then the procedure of Normalisation has to be calculated.
– Specific grain size is defined as the average size (in millimetres or centimetres) of the particles in the sample.
– Mean aperture (MA) is the opening or sieve size that would retain 50% mass of the whole sample weight.
– Coefficient of variance is the standard deviation of the particle size distribution, expressed as a percentage of the MA.
– This data can be automatically calculated and displayed by computer programmes or plotted graphically.

17.3 Particle counters

These are electronic devices that are commonly used to measure how clean the environment is in a so called cleanroom; the counters detect and count particles, such as aerosols, atomised liquid mists and minute physical particles. There are principally two types of measurement: optical and condensation.

The optical type uses a light source which is blocked or scattered when a particle passes through the light beam.

The condensation type uses an alcohol-type liquid to enlarge the image of the particle.

There is another type of counter which works on the principle of impedance or conductance changes due to change in very small particle sizes, such as blood cells, whilst passing through an aperture of a tube suspended in an electrolytic medium. This principle is known as the **Coulter principle**. The technique is also used for measuring small sizes of fat globules in suspensions in products such as cosmetics. The particle range is generally from 0.4 μm up to 1.6 mm.

17.4 Imaging

The principle is the measuring of the area and volume of the particles photometrically and computing the output.

This technique is especially useful for measuring loose or imbedded particles of about 1–25 µm in size.

There are many types and models of computerised instrumentation available on the market.

Chapter 18
Surface tension

Keywords: tangential force, types of instrument, penetrometer

The outline of this chapter is the study of forces and measurement techniques of the principles of surface attraction or interfacial tension with air (or the container surface to that of the liquid); these are utilised in the laboratory to obtain data on the depth of penetration of a weight subjected to by a material on the surface of a particular liquid over a prescribed time period.

Surface tension, by definition, is the **attraction** of a liquid's closely **packed molecules** (i.e. the cohesion of like molecules (liquid) with like molecules (liquid) compared to the cohesion of unlike moles of the air) at the surface of a liquid.

Basically, it is the **tangential force** acting at a liquid's interface with the atmospheric air. In other words, it is similar to the skin on the surface of the water which is why many insects can walk onto the surface without sinking (drowning).

This force of attraction of two bodies is sometimes referred to as **capillary force**, **surface energy**, **surface free energy** or **interfacial tension**.

Pure water is defined as having a surface tension of 1.0.

There are various types of instruments or techniques that can be used to measure this force of attraction, such as:

- **Du Nouy ring** (the common and general method), where an applied force is applied to lift a platinum ring from beneath the surface of the liquid to the surface, then to break that tension (attraction) between liquid and air.
- **Wilhelmy plate** is similar to the Du Nouy ring, but instead of a ring, it uses a thin plate which is oriented perpendicular to the surface of the liquid being measured by its adhesion tension on the vertical movement of the plate in the liquid.
- **Pendant drop** is classed as an optical technique.
- **Staglagmometer** is a very simple device also known as drop volume tensiometer), the **principle** here is to measure the weight of drops of a fluid (sample liquid) falling from a capillary glass tube (such as pipette or burette or indicator dropper bottle); then calculate the surface tension of the fluid by **Tate's law**. This law states that the mass (m) of the falling drops (attraction of a drop at the surface of the liquid) depends on the surface tension (σ) and the capillary radius (r) under the force of gravity (g):

$$mg = 2\pi r\sigma$$

This equation describes the fact that the drop falls at the exact moment in time that its weight counters the balance of the force applied to form the new liquid surface, at the breakaway limit.

https://doi.org/10.1515/9783110721201-019

This type of apparatus is used in testing emulsions (surfactants).

Another type of similar instrument is the **penetrometer which** is an instrument for measuring the consistency (firmness), strength of compaction, softness or hardness; of a soil material, paste or gel; by measuring the depth or rate of penetration of a rod or needle by a force or weight into the substance being measured. This type of instrument is used in the fields of botany, agricultural commodities such as fruit, soils, agronomy, hydraulic oils and greases.

18.1 Chemical analysis

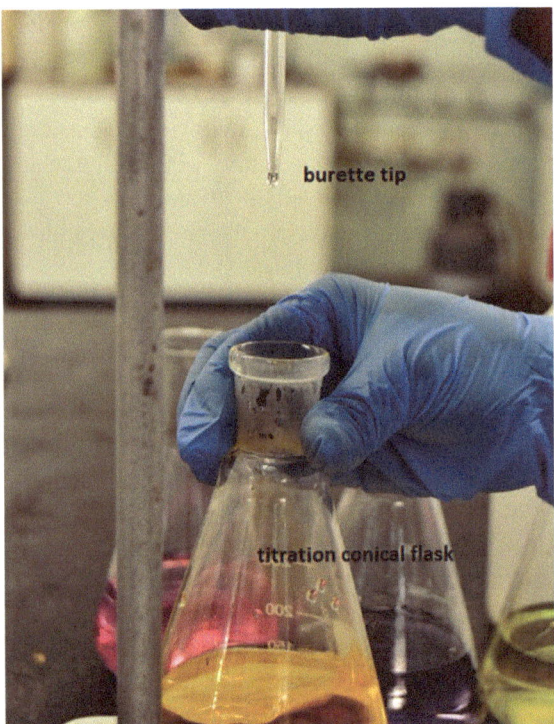

Fig. 18.1: Chemical analysis.

Chemical analysis is what analytical chemistry is all about. It involves sample preparations, wet chemistry (example would be titration, as in Fig. 18.1 above), some aspects of physical chemistry and instrumental measurements of test data controlled and recorded by the manufacturer's specific computer algorithms and analytical software programmes.

Basically, there are four types of **chemical** reactions:
- Combination (aka synthesis); $A + B = AB$
  e.g. $4Fe + 3O_2 = 2Fe_2O_3$ (rust, i.e. oxidation)
- Displacement; $A + BC = AC + B$
  e.g. $Zn + H_2SO_4 = ZnSO_4 + H_2$
- Disproportionation; $AB = A + B$
  e.g. $2H_2O_2 = 2H_2O + O_2$
- Decomposition; $AB = A + B$
  e.g. $Na_2CO_3 = Na_2O + CO_2$
- Double decomposition; $AB + CD = AD + CB$
  e.g. $NaCl + AgNO_3 = NaNO_3 + AgCl$

Common reactions are:
- Acid + base = salt + water
- Acid + alcohol = ester + water
- Halogen + metal = salt
- Acid anhydrides + water = acid
- Acid oxides + water = acid
- Metal + acid = salt + hydrogen
- Redox: $A_{oxidising\ agent} + B_{reducing\ agent} = A_{reduced\ form} + B_{oxidised\ form}$
- Combustion: $A_{combustible\ form} + O_2 = CO_2 + H_2O$ + other by-products from the combustible form
- Hydrolysis: $A^- + H_2O = HA_{aqueous} + OH^-_{base}$

Chapter 19
Chemical analysis

Keywords: electromotive series, standard electrode potentials, nomenclature acids, valency chemical reagents, preparation solutions, dilution formula, chemical bottle labels, volumetric standard solutions

19.1 Electrochemical series

The series is also known as the electromotive series. For the electrode potentials of the series, refer to Tab. 19.1 below.

Tab. 19.1: Electrochemical series.

Metal	Standard electrode potential E° (volts)
Potassium (anode; −; most active)	−2.92
Calcium	−2.87
Sodium	−2.71
Magnesium	−2.37
Aluminium	−1.66
Manganese	−1.19
Chromium	−0.90
Zinc	−0.76
Iron	−0.44
Tin	−0.14
Lead	−0.13
Hydrogen (*reference*)	**0.00**
Tin	+0.15
Bismuth	+0.2
Copper	+0.34
Iron	+0.77
Silver	+0.80
Platinum	+1.2
Gold (cathode; +; least active)	+1.5

https://doi.org/10.1515/9783110721201-020

19.2 Nomenclature of acids

Chemical analyses involves the use of correct terminology, example is naming of acids (Tab. 19.2); or the valencies of the elements and ions (Tab. 19.3 and 19.4).

Tab. 19.2: Naming of acids.

Structure of acid	Wording	Example
Most common	--------ic	HCL Hydrochloric acid
Contains more oxygen	per--------ic	HClO$_4$ Perchloric acid
Contains less oxygen	--------ous	HCLO$_2$ Chlorous acid
Contains even less oxygen	hypo----ous	HCLO **Hypo**chlor**ous** acid

19.3 Valencies

The valency or measure of combining power of a particular chemical element or ion is equal to the number of chemical bonds that one atom or ion or functional group can form when reacting with other atoms, ions or functional groups.

Tab. 19.3: Valency of an element.

Element	Atomic number	Symbol	Valency
Aluminium	13	Al	3
Arsenic	33	As	3,5
Barium	56	Ba	2
Beryllium	4	Be	2
Bismuth	83	Bi	3,5
Boron	5	B	3
Bromine	35	Br	1
Calcium	20	Ca	2
Carbon	6	C	2,4

Tab. 19.3 (continued)

Element	Atomic number	Symbol	Valency
Chlorine	17	Cl	1
Chromium	24	Cr	3,6
Cobalt	27	Co	2
Copper	29	Cu	1,2
Fluorine	9	F	1
Gold	79	Au	3
Hydrogen	1	H	1
Iodine	53	I	1
Iron	26	Fe	2, 3
Lead	82	Pb	2, 4
Lithium	3	Li	1
Magnesium	12	Mg	2
Manganese	25	Mn	2, 3, 4, 6
Mercury	80	Hg	1,2
Molybdenum	42	Mo	6
Nickel	28	Ni	2
Nitrogen	7	N	1, 2, 3
Oxygen	8	O	2
Phosphorus	15	P	3, 5
Platinum	78	Pt	2, 4
Potassium	19	K	1
Silicon	14	Si	4
Silver	47	Ag	1
Sodium	11	Na	1
Strontium	38	Sr	2
Sulphur	16	S	2, 4, 6
Tin	50	Sn	2, 4
Titanium	22	Ti	4
Uranium	92	U	6
Zinc	30	Zn	2

Tab. 19.4: Valency of an ion.

Free radical or ion	Symbol	Valency
Acetate	$C_2H_3O_2$	−1
Bicarbonate	HCO_3	−1
Bisulphate	HSO_4	−1
Carbonate	CO_3	−2
Chlorate	CLO_3	−1
Chloride	CL	−1
Chlorite	CLO_2	−1
Chromate	CrO_4	−2
Cyanide	CN	−1
Hydride	H	−1
Hydroxide	OH	−1
Hypochlorite	CLO	−1
Nitrate	NO_3	−1
Nitride	N	−3
Nitrite	NO_2	−1
Perchlorate	CLO_4	−1
Permanganate	MnO_4	−1
Phosphate	PO_4	−3
Phosphide	P	−3
Sulphide	S	−2
Sulphite	SO_3	−2
Sulphate	SO_4	−2
Thiocyanates	SCN	−1
Thiosulphate	S_2O_3	−2

19.4 Chemical reagents

In order to work precisely and accurately, the analytical chemist makes use of various commercially prepared chemical reagents dependent upon certified grades of chemical composition and purity.

It is essential that the analyst reads and understands the labels on the purchased bottles or packages. Usually, chemicals packed in amber coloured (or cobalt blue) glass bottles with ridges on the side contain poisons. The ridges indicate to a person holding the bottle that its contents are poisonous.

How to **read labels** on chemical reagent bottles:

What is a *chemical reagent bottle*?

A chemical reagent bottle is that container (glass or plastic) which contains a chemical, such as sodium carbonate, that is used for laboratory test procedures.

A laboratory worker must understand the chemical, physical and hazardous properties of the chemicals that are handled.

The label should contain the following information:
- Name of manufacturer or supplier
- Manufacturer's catalogue or product number
- Batch number
- Quantity of contents of container/bottle
- The stated grade (or purity type) of reagent
- Assay or degree of purity
- Name of chemical and its other names
- Molecular formula of chemical
- Molecular mass of chemical (if given)
- Storage conditions (if any)
- Shelf life or expiry date, or date of manufacture
- S and R phrases

Important label safety (S) *and risk* (R) *phrases*

The S and R numbers on the bottle labels refer to important Safety and Risk precautions, hazards, storage conditions, stability and necessity of any protective gear (PPE) to be worn by the laboratory worker.

These S and R phrases can be found in chemical catalogues or on the internet, and must be consulted when handling hazardous chemicals; also, these phrases explain, not only how to dispose of these chemicals or their spillages, but also how to treat any injuries.

Toxicity and flammability hazards (hazard pictograms if any), as below:

The following are the GHS/CLP published pictograms for the labelling classification of chemicals:

Oxidisers	Flammables	Explosives
	Self-reactive	Self-reactive
	Pyrophoric	Organic peroxides
	Self-heating	
	Emits flammable gas	
	Organic peroxides	

Acute toxicity (severe)	Corrosives	Gases under pressure

Carcinogen	Environmental toxicity	Irritant
Respiratory sensitiser		Dermal sensitiser
Reproductive toxicity		Acute toxicity (harmful)
Target organ toxicity		Narcotic effects
Mutagenicity		Respiratory tract
Aspiration toxicity		Irritation

Where GHS is the **globally harmonised system** and CLP is the **classification labelling and packaging** requirements for manufacturers, transporters and users of dangerous chemicals.

Note that both the words "flammable" and "inflammable" mean the same, i.e. the chemical, or substance, will burn or ignite.

The preferred word is flammable to describe something that would readily ignite or burn (combust!).

19.4.1 Preparation of some common laboratory solutions or reagents

Laboratory solutions, reagents and standards must be prepared by using distilled water or de-ionised water, or any other purified water of conductivity less than $2\,\mu S/cm$ at 25 °C.

All chemical powders or crystals must be of purity greater than 98%, such as spectrophotometric or analytical grade reagents.

When preparing standards, use must be made of an analytical balance (mass meter) capable of reading to 0.1 mg, or preferably 0.01 mg, when massing out the standard chemical.

Only calibrated grade A (class 1) volumetric glassware must be used.

Most standardised volumetric acids or bases, such as 1 normal hydrochloric acid, can be purchased already made (or in ampoule form) from various laboratory suppliers.

Where dilute acids are stated in test methods, they are generally of the following concentrations:
- Acetic acid 3 N
- Hydrochloric acid 3 N
- Nitric acid 3 N
- Phosphoric acid 9 N
- Sulphuric acid 6 N

Where dilute bases are stated in test methods, they are generally of the following concentrations:
- Ammonium hydroxide 3 N (3 M)
- Barium hydroxide 0.4 N (0.2 M)
- Calcium hydroxide 0.04 N (0.02 M)
- Potassium hydroxide 3 N (3 M)
- Sodium hydroxide 3 N (3 M)

19.4.2 Dilution formula

(also known as **Pearson's square law** of dilutions or ratios)

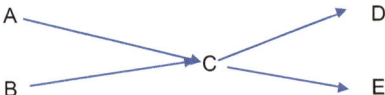

Let A (solute) and B (solvent) be the strength of the constituents in descending order of strength, respectively.

Let C be the strength required of the diluted substance or liquid.

Then E = A−C = parts of B to be taken and D = C−B = parts of A to be taken.

Example:
To prepare a 10% water (aqueous) solution of a chemical reagent X, where the concentration of chemical X is 98%:

A = 98
B = 0
C = 10

Therefore D = 10 and E = 88.

That means, take 10 parts of chemical X and 88 parts of water.
This gives a ratio of 10:88.

19.4.3 Preparation of AAS standard aqueous solutions

Generally, most atomic absorption (AA) standards are stored in 0.5 mol/L hydrochloric or nitric acids, because chlorides and/or nitrates are the most common anions found in atomic absorption spectroscopy (AAS) work (matching of sample matrix with standard matrix to eliminate spectral interferences).

Also, these acids prevent algae formation in the solutions, thus acting as preservatives. The shelf life of 1,000 ppm stock solutions is about 2 years under normal conditions of storage and use.

AA Standards should be prepared from the pure metals, however it is simpler and sometimes more accurate to prepare them from their pure metal hydrated salts, as most of these can be readily dissolved in water or dilute acids.

However, most AA standard stock solutions, such as 1,000 ppm Fe (1,000 mg/L Fe), can also be purchased ready-made (or in ampoule form) from various laboratory suppliers.

How to calculate a common series of diluted working standards required for an analysis:

A series of six diluted standards (to compile the calibration graph) may be prepared from the 1,000 ppm stock solution, as follows:

Note: Fewer standards may be prepared due to the linearity of the calibration curve; However, the standards must bracket the sample reading:

First standard	Second standard	Third standard	Fourth standard	Fifth standard	Sixth standard
A	1 × (3 × A)	2 × (4 × A)	3 × (5 × A)	4 × (6 × A)	5 × (7 × A)

where A is the lowest standard required, usually 0.5 or 0.1 ppm (or 3× the detection limit of the analyte as applicable to the type and model of the AA instrument being used).

Example: Expect sample solution to be about 6 ppm and lowest standard for AAS work to be 1 ppm Fe:

Therefore, A = 0.5; thus, the series of diluted (working) standards to be prepared will be:

(10 mL of a 1,000 ppm Fe stock standard, made up to 1 L in 0.5 mol/L hydrochloric or nitric acids; equivalent to a working standard of 100 ppm Fe):

First standard is 0.5 ppm Fe (i.e. 0.5 mL of 100 ppm standard made up to 100 mL)

Second standard is $1 \times (3 \times 0.5) = 1.5$ ppm Fe (i.e. 1.5 mL of 100 ppm standard made up to 100 mL)

Third standard is $2 \times (4 \times 0.5) = 4.0$ ppm Fe (i.e. 4.0 mL of 100 ppm standard made up to 100 mL)

Fourth standard is $3 \times (5 \times 0.5) = 7.5$ ppm Fe (i.e. 7.5 mL of 100 ppm standard made up to 100 mL)

Standard solutions fifth and sixth need not be prepared in this instance as the first four standards bracket the expected level of the sample solution.

The above procedure can also be applied to ICP calibration standards where the stock standard is usually of 10,000 ppm concentration and the solute is 1% or 2% nitric acid.

Series diluted standards are usually 0.1, 1, 10 and 100 ppm.

How to calculate the exact strength of the standard solutions prepared using primary standards in the titrations (i.e. standardisation):

For NaOH solutions (using potassium hydrogen phthalate):

$$N \, (\text{or M}) = \frac{\text{g } C_8H_5KO_4}{0.20423 \times \text{mL titration}}$$

For HCl solutions (using sodium tetraborate):

$$N \, (\text{or M}) = \frac{\text{g } Na_2B_4O_7 \cdot 10H_2O}{0.19072 \times \text{mL titration}}$$

For H$_2$SO$_4$ solutions (using sodium tetraborate):

$$N \, (\text{or M}) = \frac{\text{g } Na_2B_4O_7 \cdot 10H_2O}{0.19072 \times \text{mL titration}}$$

Refer to Tab. 19.5 for the preparation of standardised reagents.

Tab. 19.5: Preparation of 1 L of some volumetric standard solutions.

The following solutions are made up to 1 L volume with pure water:					
Standard solution	Mass chemical used for preparation of Standard	Standardised against: (primary standard)	Burette size to be used (use grade A burette only)	Indicator solution to be used	Approx. titration (example only)
0.1 N NaOH 0.1 M NaOH 0.1 mol/L NaOH	4.00 g NaOH pellets	±0.3 g potassium hydrogen phthalate (as is)	25 mL	1% phenolphthalein	12.0 mL
1 N NaOH 1 M NaOH 1 mol/L NaOH	41 g NaOH pellets	±1.0 g potassium hydrogen phthalate (as is)	10 mL micro burette	1% phenolphthalein	5.0 mL
0.1 N HCl 0.1 M HCl 0.1 mol/L HCl	9.33 mL of HCl of SG 1.16; or 8.3 mL of HCl of SG 1.19	±0.2 g sodium tetraborate decahydrate (as is)	25 mL	0.02% methyl red	12.0 mL
1 N HCl 1 M HCl 1 mol/L HCl	93.3 mL of HCl of SG 1.16; or 83 mL of HCl of SG 1.19	±1.0 g sodium tetraborate decahydrate (as is)	10 mL microburette	0.02% methyl red	5.0 mL
0.1N H_2SO_4 0.05 M H_2SO_4 0.05 mol/ L H_2SO_4	5.0 g H_2SO_4 of SG 1.84	±0.2 g sodium tetraborate decahydrate (as is)	25 mL	0.02% methyl red	12.0 mL

Chapter 20
Gravimetric analysis

Keywords: furnace temperatures, classical gravimetric analysis, co-precipitation, post-precipitation, von Weimarn rule, Fajans–Paneth–Hahn rule, silver nitrate reactions, Mohr precipitation titration

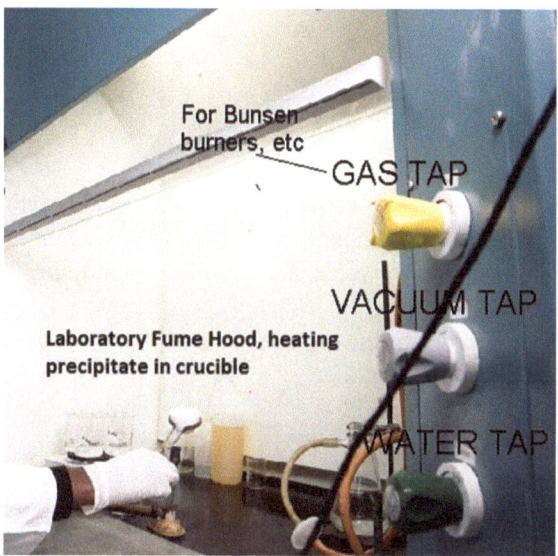

Fig. 20.1: Igniting precipitate in a fume cabinet.

Gravimetric analysis is called as such because it involves weighing (massing out) the sample and then weighing the result of the final chemical reaction of the analysis (the product, namely the precipitate); after it has been separated by filtration and the filter paper (or crucible) ignited to a final temperature on a Bunsen burner or in a Muffle furnace, usually at 950 °C.

Approximate **furnace temperatures** can be gauged by the colour of the crucible, as follows:

– Dull red = 700 °C
– Cherry red = 900 °C
– Orange = 1,100 °C
– White = 1,300 °C

The gravimetric technique as well as the technique of volumetric titrations is sometimes referred to as **classical** methods, wet chemistry or bench chemistry, as most

https://doi.org/10.1515/9783110721201-021

of the work is done at the laboratory bench and not measurement by instrumentation such as automatic titrators.

Some common uses of gravimetric analyses are:
-   Determination of moisture content of a substance by weighing before and after oven drying
-   Determination of chlorides by weighing the precipitate formed from a chemical reaction with silver nitrate
-   Determination of sulphates by weighing the precipitate formed from a chemical reaction with barium chloride
-   Determination of magnesium by weighing the organic precipitate formed from a chemical reaction with 8-hydroxyquinoline
-   Determination of nickel by weighing the organic precipitate formed from a chemical reaction with dimethylglyoxime

The **principle** of the gravimetric technique is the **separation** of an element or compound from the sample matrix by **reacting** it with another compound or reagent to form a compound that is insoluble (**precipitate**) which can be isolated and **quantified** by weighing.

The analytical requirements of the **precipitation formation** are:
-   It must be completely insoluble.
-   There must be no co-precipitation.
-   There must be no post-precipitation.
-   It must be stable.
-   It must be capable of separation from the rest of the sample components.
-   Separation must be easy, quick and complete; this can be achieved by filtration or centrifugation.
-   It must be in a weighable form of a defined chemical composition, so that accurate stoichiometric calculations can be done to determine the concentration of the analyte.
-   The chemical reactions must not form turbidity or colloidal states; e.g. the precipitate must be able to settle out gravimetrically into a pure state.

Von Weimarn rule for the size formation of the nuclei.

Optimum conditions for precipitation are the slow addition of dilute reagents with constant stirring, then allowing the precipitate to mature for several hours.

20.1 Co-precipitation

This is where the formation of the precipitate occludes or adsorbs other unwanted substances, leading to contamination of the precipitate. An ion that is similar or common to the precipitate composition will be strongly adsorbed.

The **Fajans–Paneth–Hahn rule** is where the ion most strongly adsorbed will be the one which forms the least soluble compound with an ion of the precipitate.

20.2 Post-precipitation

This is where other unwanted insoluble substances start to settle out or contaminate the precipitate due to extended time periods where the precipitate is left standing in its original reaction form. This sometimes occurs in super-saturated solutions; refer to the von Weimarn theory of crystal growth.

An excellent illustration of the principle of gravimetric analysis is the common determination of chlorides using silver nitrate reagent.

The amount of chloride ion in solution (say, in drinking water) is to be measured, recorded and reported. The chloride ion is determined by precipitation with a solution of silver nitrate, which is slightly acidified with nitric acid (or sulphuric acid) to reduce possibilities of co-precipitation with other halides or even carbonates and phosphates:

$$Na^+ + Cl^- + Ag^+NO_3^- = Ag^+Cl^- + Na^+NO_3^-$$

The water sample is slightly acidified with dilute nitric acid and the silver nitrate reagent solution is added slowly with stirring, until complete precipitation has taken place. To check if sufficient silver nitrate has been added, wait until the precipitate has somewhat settled to bottom of flask, then add the silver nitrate reagent again, but slowly down the side of the beaker without disturbing the sediment; if more precipitate starts forming, then insufficient reagent had initially been added. Leave the precipitate to mature for several hours, no longer as post-precipitation might occur and hence lead to errors.

Refer to Tab. 20.1 for possible chemical side reactions during any argentiometric analysis.

Another common technique with silver nitrate and chlorides is the **Mohr precipitation titration** (**argentometric titration**) using potassium chromate as the adsorption indicator.

An excellent illustration of the principle of gravimetric analysis is the common determination of sulphates using 10% barium chloride solution acidified with hydrochloric acid, as the precipitating reagent (a double displacement reaction). The white precipitate of barium sulphate is then weighed:

$$2Na^+ + SO_4^{2-} + Ba^{2+} + 2Cl^- = Ba^+SO_4^{2-} + 2Na^+Cl^-$$

Note that barium sulphate is only soluble in hot concentrated sulphuric acid.

Tab. 20.1: Reactions with silver nitrate.

Ion	Reactions and observations
Chlorides	White precipitate of silver chloride, soluble in ammonia but insoluble in nitric acid
Iodides	Yellowish precipitate of silver iodate, insoluble in ammonia, but soluble in sodium thiosulphate
Bromides	Pale yellow to cream precipitate of silver bromate, soluble in ammonia
Chromates	Reddish precipitate of silver chromate, soluble in dilute nitric acid
Hydroxides	White precipitate turns to brown silver hydroxide, soluble in dilute nitric acid
Carbonates	Yellowish precipitate of silver carbonate, soluble in dilute nitric acid
Thiocyanates	Insoluble in nitric acid

Chapter 21
Volumetric analysis (titrimetric)

Keywords: titrations, types of reactions, Karl Fischer volumetric titration, Hydranal reagents, indicators, coulometric titrations, argentometric titrations, pipette, burette, burette reading, acid conversion factors, types of titrations

This chapter deals with the common wet chemistry techniques and chemical reactions of titrations.

Fig. 21.1: Titrating acid to base reaction.

The **principle** here is the volumetric measurement of the sample in solution, reacting with added reagents, to produce a liquid product (or precipitate) of the analyte. The analyte in solution (or as a precipitate) is then converted by a specific chemical equation into a quantifiable level of concentration of the analyte in the sample.

The skilful use of volumetric flasks, pipettes and burettes is critical for accurate and precise test results. Note the analyst titrating using a burette, in Fig. 21.1.

A laboratory worker should be able to manually undertake simple titrations using a burette, although many laboratories have automated titrating apparatus connected to autosamplers and controlled by computer programs; examples are those manufactured by Metrohm, Mettler Toledo and Sartorius.

https://doi.org/10.1515/9783110721201-022

21.1 Titrations

Basically, titrations are reactions between two or more reactants (chemical compounds) to produce one or more equimolar products (chemical compounds), using a colour dye indicator, by precipitation or by electrochemical sensors (pH – potentiometry or redox) to determine the equivalence or end point (completion) of the stoichiometric reaction.

Titration analysis (aka **volumetric** analysis) is one of the oldest quantitative techniques used in analytical chemistry; the other old technique is gravimetric analysis.

Gravimetric analysis is similar to that of volumetric analysis, except the product of the reaction (**precipitate**) is weighed to determine, by stoichiometric calculations, the level of the analyte in the sample.

Most titrimetric determinations are determined in aqueous (water) media. However, sometimes the titration is required to be undertaken in a non-aqueous media such as total base number of mineral oils in glacial acetic acid or acetic anhydride media.

The main **disadvantages** of titrimetry are: (i) the volumes of accurately standardised solutions are required and some reagents are unstable, such as the indicator dyes (powders or solutions), and (ii) the skilled analysts should not be colour blind so that they can work pedantically under scrupulously clean glassware apparatus.

Note the use of PPE and white apparel

Fig. 21.2: Adjusting the height of burette for volumetric analysis.

Titrations are basically very simple measurements of **two volumes** (the standardised solution of known concentration, that is, the titrant, usually in the burette) and the resulting titer (aka titre), that is, the burette reading (usually the sample or analyte is in the conical flask). The substance being titrated is known as the titrate. Titrations can be done manually by inexperienced persons (note that the analyst is checking, not reading, the level of titrant in the burette during a titration process, as shown in Fir, 21.2 above.) or automatically by computerised titrators and autodispensers.

Common oxidation titrations are:
- potassium permanganate: a very strong oxidising agent in the presence of sulphuric acid, standardised against ferrous ammonium sulphate or sodium oxalate;
- potassium dichromate: a more stable solution than potassium permanganate which is subject to organic/dust contamination; also can be standardised against ferrous ammonium sulphate;
- ceric sulphate: an expensive reagent;
- iodine;
- potassium iodate;
- potassium bromate.

Common reduction titrations are:
- sodium thiosulphate and
- arsenious oxide.

The main problem with manual titration is the degree of colour blindness of the analyst to detect the exact end point (colour change of the indicator).

Titrations are usually undertaken when the concentration of the analyte is not lower than 0.001%, but can be up to 100% which can then be diluted to a suitable measurement level. Precision of the analysis is dependent upon the skill of the analyst and can be better than 1%. Generally, the precision of any manual analytical process, such as fat extraction and determination, is not better than 5%.

Titration methodology varies from simple inorganic ions, such as determining water hardness (calcium and magnesium salts), to complex organic analysis, such as vitamin C determination using indophenol.

Basic types of reactions in titrimetry:
- **neutralisation** reaction, e.g. acid–base titration;
- **conductometric** titrations (end point is difficult to discern, unless an automatic titrator has been used);
- **potentiometric** titrations using pH or ion-selective electrodes;
- **complexometric** titrations, e.g. EDTA titration for alkali and alkaline earth ions;
- **precipitation** titration, e.g. silver chloride;
- **argentometric** titration e.g. silver nitrate;

- **amperometric** or **voltammetric** titrations, e.g. determining the different species of chlorine using dual polarisable electrodes measuring currents at a constant voltage (dropping mercury electrode);
- **potentiometric volumetric** titration, namely the Karl Fischer (KF) volumetric titration for determining water content of various materials, using Hydranal-Composite® or Hydranal-Titrant/Solvent®. Note that there are other makes and generics of KF titration (KFT) reagents on the market, other than Hydranal®, such as Honeywell, Fluka and Aquastar.

Selectivity is critical when the analysis (titration) is to be done on a complex sample matrix. There are various procedures to ensure that the analyte being titrated is free of interfering substances or side reactions. A few of these precautions or pre-treatments are:
- complexation; e.g. EDTA titrations where pH adjustment will remove interfering ions from yielding similar endpoints;
- back-titrations by adding excess titrant, e.g. determination of chlorides with silver nitrate in unknown sample matrices;
- precipitation by removing first an interfering analyte;
- pH adjustments by addition of buffers;
- selective extraction;
- ion exchange to remove interfering cations or anions.

The object of **KFTs** is the determination of moisture in a sample or the water of hydration (water of crystallisation) of a salt, where the substance would degrade if any heat is applied to it, such as moisture by oven drying or by infrared lamps.

The basic principle is that water reacts with iodine and sulphur dioxide in the presence of a base and alcohol. In the coulometric technique, iodine is electrolytically generated in a reagent (known as the anolyte) which contains iodide.

KF volumetric titration is universally used for determining micrograms of moisture present in all types of sample matrices, even for determining the purity of water samples. Various Hydranal™ reagents or similar suppliers are required. Note that Hydranal™ reagents are pyridine free of this toxic and noxious additive.

Headspace (HS) volumetric KFTs are used where there can be physicochemical interferences present from the matrix of the sample (e.g. tobacco products). Here the sample is oven dried in gas chromatography vials (septum capped) and the vapour injected (e.g. via syringe) into the KF titrating cell, to react with the water molecules present.

Some instances where HS KF is useful, are listed below:
- samples does not dissolve in the Hydranal® reagents;
- poor reproducibility;
- no end point can be reached, because the sample components, other than water, are reacting with the KF reagents;
- samples that only release their water content at high temperatures;
- the sample forms stubborn residues on the electrodes and the titrating cell.

The **main** reaction of KF is the conversion of alkylsulphite by oxidation to alkylsulphate in the presence of iodine. The water in the sample reacts with the iodine in the Hydranal reagent in a stoichiometric ratio which causes a potential voltage drop across the electrode pair of the instrument. This millivoltage is measured and recorded by the titrating instrument, or manually iodometrically with a burette and is proportional to the concentration of the water present in the sample.

Hydranal-Composite is the most common reagent for a one-component titration, it contains iodine, sulphur dioxide, bases and stabilisers; note that these reagents are very sensitive and reactive in the presence of humidity in the laboratory. This reagent has unlimited water capacity (up to 100%) and is stable to about 5 years.

The solvents in these reagents are mostly methanol; however, other solvent media may be necessary depending on the solubility properties of the sample matrix.

Some examples of reagents and their uses are:
- Hydranal-Composite 1, which has lowest sensitivity of 1 mg water per mL reagent;
- Hydranal-Composite 2, which has a sensitivity of 2 mg water per mL reagent;
- Hydranal-Composite 5, which has a sensitivity of 5 mg water per mL reagent;
- Hydranal-E media used ethanol, in place of the toxic methanol;
- Hydranal-K media used chloroform, for titration of ketones and aldehydes;
- Hydranal-Solver is suitable for non-polar samples such as oils and fats, which has a poor solubility in methanol.

Hydranal-Titrant/Solvent is the most common reagent for a two-component titration. Here, KF reagents are separated into two solutions, namely, the titrant contains a known concentration of iodine (which is in the burette) and the solvent that contains sulphur dioxide and a base, as well as the sample, is in the titration flask.

The two-component reagent is suitable when measuring milligram amounts of water in a sample, where a high buffering capacity is required when titrating acidic samples. This reagent has a shelf life of 2 years (titrant) and 5 years (solvent).

21.2 Indicators

What is an *indicator*?

An indicator is a natural or synthetic substance that changes colour in response to the nature (electronic charges) of its chemical environment. Indicators are used to provide information about the degree of acidity/alkalinity of a substance (pH) or the state of some chemical reaction within a solution being tested or analysed.

One of the oldest indicators is **litmus**, a lichen-derived dye that turns red in acid solutions and blue in basic (alkaline) solutions. Other common indicators include alizarin, methyl red, methyl orange and phenolphthalein, as well as some mixed indicators such as xylene cyanol FF such as those shown in Tab. 21.1 below, each one being useful for a particular range of acidity or a certain type of chemical reaction.

pH indicators must be selected such that their pK value is closest to the pH value of the end point of the titration.

Mixed indicators are used commonly in routine testing laboratories for various reasons.

Tab. 21.1: Mixed (screened) indicators.

Indicator	pH range	Colour change	Preparation
Screened methyl orange	2.9–4.4	Magenta to grey to green	Equal volume of 0.2% methyl orange + 0.3% xylene cyanol FF, in 50% alcohol
Screened methyl red	4.2–6.3	Magenta to grey to green	Equal volume of 0.2% alc. methyl red + 0.1% aq. methylene blue
Bromocresol green/methyl red	4.2–5.2	Pink to grey to green	Equal volume of 0.15% alc. bromocresol green + 0.1% alc. methyl red

21.3 Coulometric titrations

The principle is where an electrode is generated from chemical reagents by electrolysis (an applied voltage potential between an electrode and its generated paired electrode).

The most common type is the coulometric KFTs for low moisture determinations.

Here the solvent used to disperse the sample which could be either a liquid or solid is critical, usually it is methanol based. The complex titrant, usually Hydranal-Coulomat, is important in deciding which to use depending upon the matrix of the sample.

The coulometric KFTs require two types of reagent solutions: an **anolyte** (solution) and a **catholyte** (electrode):

- The Hydranal-Coulomat A-type for cells with diaphragm or type AG with or without diaphragm; used as the anolytes.
- The Hydranal-Coulomat CG-type used as the catholytes.
- The Hydranal-Coulomat E-type used as both anolytes and catholytes.
- The Hydranal-Coulomat F-type used as the anolytes.
- The Hydranal-Coulomat Oil have blends of methanol, chloroform and xylene as solvents for water in oil determination.
- The Hydranal-Coulomat reagents are based on different solvent blends depending upon the sample matrix.

The other type of KFT is the volumetric technique for water contents from 1 ppm to 99.99% concentration.

There are generally **two types** of coulometric KF titrators, those with a **diaphragm** and those without one. The diaphragm separates the anode part from the cathode part of the coulometric cell. Oxidation of the iodine ion to iodine takes place at the anode and the reduction of the hydrogen ions (in the water) to hydrogen at the cathode.

For those titrators that are without a diaphragm, only one reagent is required, the anolyte. However, these are not as accurate as those with a separating diaphragm, which can detect much lower concentrations of water in micrograms.

21.4 Argentometric titrations

A common technique with silver nitrate and chlorides is the **Mohr precipitation titration** using a 5% potassium chromate solution as indicator. The pH of the sample must be adjusted between 6.5 and 9. The precipitate is dark orange-red just before a brownish colour appears.

Another common precipitation technique with silver nitrate and chlorides is Fajan's titration using fluorescein (a greenish suspension dye) as an adsorption indicator. The pH of the sample must be adjusted between 5 and 8.5. The precipitate is a pinkish colour. For bromides, iodides and thiocyanate ions, use is made of eosin as the indicator.

Another common precipitation technique with the addition of excess silver nitrate reagent and reactions with chlorides (as the analyte) is the **Volhard's back titration**. The excess silver nitrate (i.e. that amount was not consumed in the reaction with chlorides) is removed by filtration, and the amount of silver nitrate in the filtrate is determined by titration with ammonium thiocyanate using ferric ammonium sulphate as indicator. The pH of the sample must be acidic. The end point is a distinct red colour.

Note that silver halide compounds are light sensitive.

Volumetric (volumetric flasks) and graduated glassware (burette and pipettes) are usually graded as either Grade A (Class A, Class 1) or Grade B (Class B, Class 2) glassware.

Grade A glassware is individually batch checked and calibrated at the glass factory, and a certificate of compliance or calibration (with a serial number for traceability) should be issued by the manufacturer. Class A pipettes (excluding the blow-out type) are also marked with the delivery time in seconds, e.g. 20 s.

Grade B glassware is only randomly quality checked per batch (at the manufacturer), and this type of glassware is generally cheaper than Grade A and less accurate.

A **pipette** is a graduated tube, with or without a scale (i.e. graduated or bulb), used for transferring pre-measured or fixed volumes of liquid. When handling toxic or corrosive liquids, always use a safety pipettor. (It is important to abide by the drainage times marked on this glass apparatus.)

There are various types of pipettes available, which require different delivery procedures:

Blow-out pipettes (where last retaining drop is blown out); marked BLOW OUT at the top of the pipette;
- delivery from 0 at top, to tip (jet);
- delivery from 0 at top, to graduation mark at bottom, not to tip (jet);
- delivery from a graduation mark at top (e.g. 25 mL) to 0 mark before the tip (jet).

A **burette** is a graduated glass tube with a tap (stopcock) and a fine outlet tube (jet) used for delivering or dispensing known accurate volumes of liquid, e.g. used in titrations.

See below (Fig. 21.3), on how to use and read a burette. (It is important to abide by the drainage times marked on this glass apparatus.)

Volumetric ware is graduated in millilitres (mL) in accordance with BSI, DIN or ISO Standards. Some micro-pipettes are graduated in lambda units (λ), where $1\lambda = 0.001$ mL. Most volumetric ware (pipettes, burettes and volumetric flasks) are calibrated for use at 20 °C (i.e. the volume that it holds is only exact (accurate) when the liquid and the volumetric ware are both at 20 °C (because volume of a liquid is, of course, dependent upon its temperature). The markings on these volumetric glassware will state the temperature, tolerance, brand and the specification that they were manufactured to.

21.5 Neutralisation titrations (acids and bases)

A common chemical procedure where an acid solution is titrated (reacted) against an alkali solution or vice versa, using a colour dye to determine the equivalence point (stoichiometric end point), to produce a salt plus water as products of the reaction.

Some **acid conversion factors**, commonly used in food quality control laboratories, where titrant used 1 mL N/1 (or 1 M) NaOH solution, is equivalent to:
- 0.04603 g HCOOH (formic acid)
- 0.09008 g $C_3H_6O_3$ (lactic acid)
- 0.03646 g HCl (hydrochloric acid)
- 0.04904 g H_2SO_4 (sulphuric acid)
- 0.06005 g CH_3COOH (acetic acid)
- 0.07005 g $C_6H_8O_7 \cdot H_2O$ (citric acid monohydrate)
- 0.0640 g $C_6H_8O_7$ (citric acid anhydrous)
- 0.05804 g $C_4H_4O_4$ (maleic acid)
- 0.06706 g $C_4H_6O_5$ (malic acid)
- 0.28245 g $C_{18}H_{34}O_2$ (oleic acid (for tallows and fish oil))
- 0.07504 g $C_4H_6O_6$ (tartaric acid)
- 0.256 g $C_{16}H_{32}O_2$ (palmitic (for palm oil))
- 0.200 g $C_{12}H_{24}O_2$ (lauric (for crude and refined palm kernel and coconut oils))

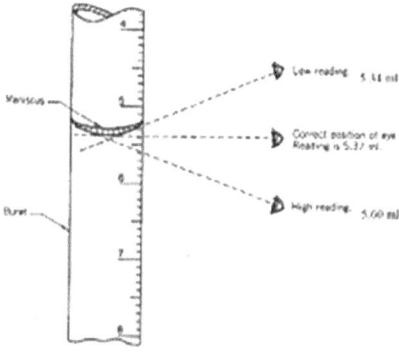

Diagram showing the correct way to titrate using a burette.

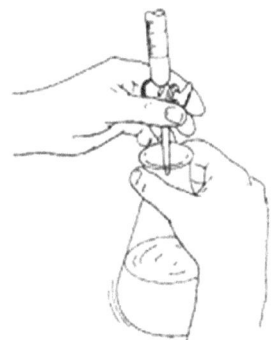

Fig. 21.3: How to read a burette.

- 0.12210 g C_6H_5COOH (benzoic acid)
- 0.06184 g H_3BO_3 (boric acid)
- 0.06303 g $C_2H_2O_4$ (oxalic acid)
- 0.04900 g H_3PO_4 (o-phosphoric acid)
- 0.17613 g $C_6H_8O_6$ (L-ascorbic acid)

For example:
- fruit juices use citric acid factor;
- grapes use tartaric acid;
- peaches, apricots and plums use malic acid.

Example of a calculation:
% free fatty acids, as oleic acid = titration mL × N NaOH × 0.28245 × 100 wt sample in g

21.6 Indicators for Neutralisation titrations (acids and bases)

This is a common chemical procedure where an acid solution is titrated (reacted) against an alkali solution or vice versa, using a colour dye (indicator, such as those stated in Tab. 21.2 below) to determine the equivalence point (stoichiometric end point), to produce a salt plus water as products of the reaction.

Tab. 21.2: Indicators for (neutralisation) acid–base titrations.

pH indicator	pH value	pK$_a$ value	Colour change	Preparation
Alizarin	5.5–6.8	6.9	Yellow to red	0.1% in methanol
Bromophenol blue	3.0–4.6	4.1	Yellow to blue-purple	0.04% in 20% alcohol
Bromocresol green	3.8–5.4	4.9	Yellow to blue-violet	0.1% in alcohol
Bromocresol purple	5.2–6.8	6.3	Yellow to purple	0.04% in 20% alcohol
Bromothymol blue	6.0–7.6	7.3	Yellow to blue	0.04% in 20% alcohol
Cresol red	0.4–1.8 7.0–8.8	1.0	Red to yellow Yellow to red	0.1 g in 26 mL 0.01 M NaOH + 224 mL water
Congo red	3.0–5.0	4.5	Blue to scarlet	0.1% aqueous
Litmus	4.5–8.0 7.5–14.0		In acid, purple turns to red In alkali, red turns to purple	Litmus paper
Methyl orange	3.0–4.4	3.5	Red to orange-yellow	0.1% aqueous
Methyl red	4.2–6.3	4.8	Red to yellow	0.2% in alcohol
Neutral red	6.8–8.0	6.8	Red to orange	0.01 g in 50 mL alcohol + 50 mL water
Phenolphthalein	8.2–9.8	9.5	Colourless to red	1.0 g in 100 mL of 60% alcohol
Phenol red	6.7–8.2	7.9	Yellow to red	0.05% in 20% alcohol
Thymol blue	1.2–2.8 8.0–9.6	1.7 9.2	Red to yellow Yellow to blue	0.04% in 20% alcohol
Thymolphthalein	9.4–10.4	9.9	Colourless to blue	0.04 g in 50 mL alcohol + 50 mL water

For titrations of strong acid (such as hydrochloric acid or sulphuric acid) versus strong base (such as sodium hydroxide):
- For 1 normal solution of strong acid and strong base use any indicator which has a pH range between limits 1 and 11 (e.g. methyl orange)
- For 0.1 normal solutions of strong acid and strong base use an indicator which has a pH range between limits 4.5 and 9.5 (e.g. methyl red or phenolphthalein)

– For 0.01 normal solutions of strong acid and strong base use an indicator which has a narrower pH range, hence between limits 5.5 and 8.5 (e.g. methyl red, phenolphthalein, bromothymol blue; or phenol red in absence of dissolved carbon dioxide gas).

For titrations of weak acid (e.g. acetic acid) versus strong base (e.g. potassium hydroxide) where the pH range is usually between 8 and 10.5, use indicators such as thymol blue, phenolphthalein or thymolphthalein.

Note: for coloured solutions/samples, indicator dyes cannot be used as the end point cannot be seen, thus potentiometric titrations (using pH meter or other suitable instruments) can be done, by using indicator electrodes, such as pH, ORP, chloride or other specific ion electrodes.

21.7 Redox (reduction–oxidation) titrations

Redox titrations are basically the exchange of the outermost shell electrons between two ionic compounds.

One compound is reduced to a lower valence oxidation state by gaining an electron and the other compound is oxidised to a higher valence oxidation state by losing an electron. See Tab. 21.3 and 21.4 for the reagents and indicators usually used in these type titrations.

Tab. 21.3: Redox reagents.

Reagent solution	Most common concentration in mol/L
Ammonium ceric sulphate	0.10
Ceric sulphate	0.10
Ferroin	0.025
Iodine	0.50
Potassium bromate-bromide	0.0167
Potassium dichromate	0.0167
Potassium iodate	0.05
Potassium permanganate	0.20
Sodium arsenite	0.05
Sodium thiosulphate	0.10

Tab. 21.4: Redox indicators for reduction–oxidation titrations.

Indicator	Redox potential @ 20 °C, pH 7	Colour change: oxidised form to reduced form	Preparation
Barium diphenylamine-4-sulphonate	+0.8	Red to colourless	0.005 M solution in water
Diphenylamine	+0.8	Blue to colourless	Dissolve 1 g in 100 mL conc. sulphuric acid
Ferroin (1,10-phenanthroline ferrous complex)	+1.1	Blue to Congo red	Dissolve 0.7 g ferrous sulphate + 1.5 g 1,10-phenanthroline in 100 mL water
Methylene blue	+0.5	Blue to colourless	Dissolve 0.2 g in 100 mL water
Neutral red	−0.3	Violet-red to colourless	Dissolve 0.5 g in 100 mL alcohol
Indophenol (2,6-dichlorophenol)	+0.2	Blue to colourless	Dissolve 0.2 g sodium salt dehydrate in 100 mL water
Safranin T	−0.3	Blue violet/brown to colourless	Dissolve 0.5 g in 100 mL water
Starch	Specific in iodine titrations	Blue to colourless	Add 1 g soluble potato starch, wet it into a paste, then add with stirring, into 100 mL boiling water, and then cool
Xylene cyanol FF	+1.1	Yellow to pink	Purchase locally

21.8 Precipitation titrations

Example: In argentometry, the titrant is silver, Ag^+ (silver nitrate):

$$Cl^- + Ag^+ = AgCl \text{ precipitate}$$

An example of a calculation for determination of chlorides in water (sample water must be between pH 7 and 10; to remove errors in the precipitation of the chromate ion), using potassium chromate (Mohr method) as indicator (for other types indicators used in these type titrations, refer to Tab. 21.5 below) is as follows:

$$1\,mL \text{ of } 0.01\,M \text{ AgNO}_3 \equiv 0.3546 mg \text{ Cl}^-$$

$$mg \text{ Cl}^-/\text{Litre} = \frac{(A - B) \times N \times 35.46\,mg \times 1,000}{1 \times mL \text{ sample}}$$

where;

$$A = mL \text{ titration for sample}$$
$$B = mL \text{ titration for blank } (= 0.2 \, mL)$$
$$N \text{ or } M = \text{normality of } AgNO_3 \text{ solution used}$$

Tab. 21.5: Adsorption indicators for precipitation titrations.

Indicator	Colour change	Preparation
Alizarin red S	Yellow to pink	0.2 g sodium alizarin sulfonate in 100 mL water
Bromophenol blue	Colourless to blue	Dissolve 0.1 g in 100 mL water
Fluorescein	Yellow-green to red	Dissolve 0.2 g in 100 mL 70% alcohol
2,7-Dichlorofluorescein	Yellowish to pink	Dissolve 1 g in 1 L 70% alcohol
Eosin	Pink to reddish	Dissolve 0.2 g in 100 mL 70% alcohol
Potassium chromate	Yellow to orange brown	4.2 g pot. chromate + 0.7 g pot. dichromate in 100 mL water
Rose Bengal	Pink to bluish	Buy commercial
Tartrazine	Colourless to yellow	Buy commercial

For conversions to other salts in the water sample:

$$mg \, NaCl/L = mgCl^-/L \times 1.648$$
$$mg \, KCl/L = mgCl^-/L \times 2.103$$
$$mg \, CaCl_2/L = mgCl^-/L \times 1.565$$

21.9 Complexometric titrations

Basically, this is a titration based on the formation of coordination compounds (complexes) between a metal ion and the chelating (or attaching ligand) ion. Refer to Tab. 21.6 below for some of the indicators used in these type titrations.

Examples:

EDTA:

$$Ca^{2+} + H_2EDTA^{2-} = CaEDTA^{2-} + 2H^+$$

Liebig method:

$$2CN^- + Ag^+ = Ag(CN)_2^{2-}$$

Tab. 21.6: Metal indicators for complexometric (EDTA) titrations.

Indicator	Colour change	Preparation
Calmagite	Blue to red	Dissolve 0.5 g in 100 mL water
Calcein (fluorescein)	–	Refer to Tab. 35
Eriochrome Black T	Blue to red	Dissolve 0.2 g in 15 mL ethanolamine + 5 mL alcohol
Murexide	Blue	Suspend 0.5 g in water
Patton and Reeder's reagent (HHSNNA)	Wine red to blue	Purchase commercial product
Xylenol orange	Yellow to red	Dissolve 0.5 g in 100 mL water
Thorin solution	Yellow to pink	0.2% w/v

Chapter 22
Reactions with iodine

Keywords: iodine, iodimetry, iodometry, iodide, starch indicator solution

The outline of this chapter is the chemical reactions and use of iodine solutions for measuring the levels of presence of analyte compounds.

Strong **oxidising agents** will react with iodide ion, oxidising it to **iodine**: iodometry.

Strong **reducing agents** will react with iodine, reducing it to **iodide ion**: iodimetry.

Starch is used, somewhat as an **indicator**, in iodine titrations to enhance the colour of the presence of traces of iodine. It forms a dark blue complex with iodine. An extra small amount of the starch should be added just before the expected end point of the titration as at high iodine concentrations the starch complex is too stable to show any change in colour.

To **prepare a fresh 1% starch indicator** solution, make a paste with 0.5 g of soluble potato starch with a small amount of cold distilled water, triturate, then add the paste slowly to 50 mL boiling water, whilst stirring. Boil for about a minute and then cool. This solution could last for several weeks if 0.06 g salicylic acid is added and then stored at 4 °C (refrigerator); otherwise, it is only stable for couple of days.

22.1 Iodimetry

Iodimetry is the study and practice of where a standardised **solution of iodine** is used directly in the titration of a strong reductant in a neutral or slightly acidic medium.

The end point of this direct titration is indicated by the very first appearance of free iodine being used as the titrant.

Note that iodine is more readily soluble in the presence of **iodide** (potassium iodide); the iodide suppresses the volatility of iodine. Iodine crystals are subject to sublimation.

22.2 Iodometry

Iodometry refers to the titration of iodine liberated in chemical reactions (redox):

$$I_2 \text{ (solid)} + 2\text{electrons} = 2I^-$$

The E° potential is −0.536 V.

https://doi.org/10.1515/9783110721201-023

This is the practice where excess iodine (iodide) is used in an indirect titration of oxidising agents. The iodine liberated is titrated usually with freshly standardised sodium thiosulphate solution in an acidic medium. Note that sodium thiosulphate solutions are not stable and need to be prepared fresh before use, and alternatively a trace of sodium carbonate can be added as a somewhat antibacterial or stabiliser. An alternative to sodium thiosulphate is sodium arsenite (a poisonous chemical!).

The end point of the titration is indicated by the disappearance of free iodine being titrated.

Starch is used as the indicator.

An **example** of an iodometric titration is the determination of chlorine (free) in a liquid sample, by adding an amount of iodate-free **potassium iodide**, water and glacial acetic acid. The liberated iodine is titrated with standardised solution of **sodium thiosulphate** (usually 0.1 N) using about 2 mL starch solution as indicator. The solution has a pale yellow colour and titration continued drop-wise, until the blue colour **just** disappears to colourless; this is the end point. Note that the blue colour will re-appear after a few minutes, which is normal, so **do not** continue the titration.

Note that these types of titrations must be **done quickly** to mitigate the problem of **oxidation** of the excess iodide exposure to the atmosphere, to form water and iodine. The specific iodine flask which has a tray around the neck of the flask has to be used; this tray should be wetted with the reagent to prevent atmospheric oxidation during the timed stage of reaction, before titrating.

Errors in iodometric methods:
- Early decomposition of the standardised thiosulphate solution, due to the presence of atmospheric oxygen and sunlight
- Alteration of the stoichiometric relationship between iodine and thiosulphate ion, in the presence of base
- Premature addition of starch during titration, due to the fact that starch is partially decomposed in the presence of a large excess of iodine.

Chapter 23
Qualitative chemical analysis

Keywords: qualitative, flame colours, flame zones, solubility compounds, acid radicals, spot tests, semi-microanalysis

This section deals with **microchemical** tests (using spot tests or ready-made kits) and flame tests (bead and colour identification). Colour reagent identification tests as shown in Fig.23.1 below.

Fig. 23.1: Qualitative analysis by colour.

The analytical chemist must know how to do quick and simple spot tests for the presence and identification of single element ions and separate chemical groupings. Qualitative testing is simply checking, by using simple flame colour tests (as shown in Tab.23.1), colour chart kits, field tests or complex chemical reactions (Tab. 23.2), for detecting the presence of individual elements and compounds.

Tab. 23.1: Flame colours.

Carmine red:
Lithium compounds
Crimson red:
Strontium compounds
Yellow-red (brick red):
Calcium compounds
Yellow:
Sodium compounds, some gold and iron compounds

https://doi.org/10.1515/9783110721201-024

Tab. 23.1 (continued)

White:
Magnesium

White-green:
Zinc

Green:
Copper compounds (other than halides), thallium, borates, antimony and ammonium compounds

Blue-green:
Phosphates, copper

Yellow-green:
Barium, manganese(II) and molybdenum

Blue (livid blue):
Lead, selenium, bismuth, caesium, copper, antimony and arsenic

Greenish blue:
Copper bromide and antimony

Violet-purple:
Potassium compounds

There are various chemical kits, such as test strips, on the market that do just that; however, the chemist needs to be skilled in the practice of systematic analysis of an inorganic sample or even some important organic compounds.

The **flame of a gas burner** (Bunsen or Meker burner) has basically four regions (Fig. 23.2):
– The upper outer non-luminous (colourless) region
– Outer diffusion region, upper cone
– Interconal gas region of unburnt gases, middle region light blue, **oxidising** region
– Bottom smaller inner blue cone, consisting of the mixture of air and gas (i.e. the **reducing** region); the hottest part is the blue **tip** of this bottom cone
– Bottom of the flame (i.e. the top of the chimney tube) is the coolest region

The Meker or Meker–Fisher burner has its top covered with a metal-perforated plate or mesh which acts to disperse a wider flame portion but also acts as a **flashback arrester** to prevent explosions or flame travelling back into the gas supply!

The flame temperatures are very much dependent on the air–gas mixtures.

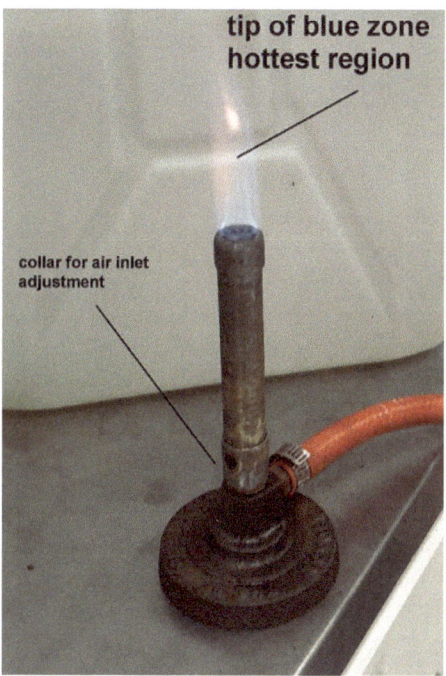

Fig. 23.2: Correct flame profile.

Tab. 23.2: Chemical reagents for spot tests.

Chemical reagent	Detection of:
Acridine hydrochloride	Co, Zn, ferric ion
Aluminon	Al
Arsenazo	Th
Benzoin α-oxime	Cu, V
α-Benzoin-oxime (cupron)	Cu, Mo
Brucine	Bi, nitrates
Cacotheline	Sn
Chromotropic acid sodium salt	Cr, Ti, nitrates
Di(4-dimethylaminophenyl)methane	Mn, I
Di(2-hydroxyphenylimono)ethane	Cd, Ca

Tab. 23.2 (continued)

Chemical reagent	Detection of:
Dimethylaminobenzylidenerhodanine	Ag, Hg, Au, cuprous ion
Dimethylglyoxime	Ni, Pd, Bi, Co, ferrous ion
2,4-Dinitro-phenyl-hydrazine	Aldehydes, ketones, vitamin C
1,5-Diphenylcarbazide	Cd, Cr
1,5-Diphenylcarbazone	Hg, V, Cd, Zn
Diphenylthiocarbazone (dithizone)	Pb
Dithio-oxamide (rubeanic acid)	Ni, Co, Cu, Ru
Dithizone	Pb, Cu, Hg, Zn
Gallocyanine	Pb
Hexanitrodiphenylamine	K, Tl
8-Hydroxy-7-iodoquinoline-5-sulphonic acid (ferron)	Ca, K, ferric ion
Hydroquinone	Ag, phosphates
8-Hydroxyquinoline	U, V
o-Nitrobenzaldehyde	Acetone, IPA
4-(4-Nitrophenylazo)resorcinol (magneson)	Mg
1-Nitroso-2-naphthol	Co
Phenylfluorone	Ge
Picrolonic acid	Ca
Quinalizarin	Be, Mg, Al, B
Rhodamine B	Sb, Ga, Au, Tl, U
Rhodizonic acid sodium salt	Ba, Ca, Pb, Sr
Salicylaldoxime	Cu, Ni, Bi, Zn
Sodium diethyldithiocarbamate	Cu
Salicylaldehyde oxime	Cu
1,2,5,8-Tetrahydroxyanthraquinone	Be, Mg, Al, In, B
Thiourea	Bi, Ru, Os
Titan yellow	Mg

Thus, a **yellowish** flame would indicate **insufficient air**, and the bottom air inlet needs to be further opened! A **blue** flame indicates sufficient air-to-gas ratio leading to hotter temperatures.

Note that all chemical reagents should be treated as being POISONOUS!

Most of the above reactions will yield colours and/or precipitates.

Some solids or salts can be identified by their crystalline structure.

The seven different types of **crystal structures** are:

- **Cubic**, a simple square box-like shape
- **Hexagonal**, a six-sided hexagon shape
- **Monoclinic**, a prism diamond-like shape
- **Orthorhombi**c, like many combined prisms
- **Tetragonal**, rectangular shape, like strips of wood splinters
- **Trigonal**, rectangular with triangular ends

The four basic varieties of solids are based on their **atomic and bonding** characteristics, namely:

- **Covalent**, e.g. organic compounds
- **Ionic**, e.g. inorganic compounds such as salt
- **Metallic**, e.g. iron, gold and copper
- **Molecular**, e.g. ammonia and water

The analyst must check on the Internet for the latest information on the above reagents for any interferences with other elements or compounds.

Further to the above simple tests, a **semi-micro** testing scheme can be undertaken by treating an aqueous solution of the test item, with a succession of various chemical reagents (Tab. 23.4 and 23.5), followed by separations by filtration, or centrifuging out precipitates.

Another common technique for quick identification of substances is the scheme for solubility tests (Tab. 23.3).

Tab. 23.3: Solubility compounds.

Compounds (inorganic)	Solubility (at room temperature)
Sodium, potassium, ammonium compounds	All soluble
Nitrates	All soluble
Chlorides	All soluble, except silver chloride, lead chloride (sparingly soluble), mercurous chloride and cuprous chloride
Sulphates	All soluble, except barium sulphate, lead sulphate, calcium sulphate and strontium sulphate

Tab. 23.3 (continued)

Compounds (inorganic)	Solubility (at room temperature)
Sulphides	All insoluble, except sodium sulphide, potassium sulphide, ammonium sulphide, barium sulphide and calcium sulphide
Carbonates	All insoluble, except sodium carbonate, potassium carbonate and ammonium carbonate
Bicarbonates	All soluble
Oxides and hydroxides	Sodium, potassium, calcium and barium are slightly soluble
Plastics	Float on water (less dense than water of SG 1.0)
Organic dirt	Float on water
Inorganic dirt or mineral matter	Sink in water
Glass	Sink in water

Tab. 23.4: Acid radical (anion) quick tests.

Reaction with	Observations
Dilute sulphuric acid	
Carbonates (CO_3^{2-})	CO_2 gas evolved with effervescence, turns lime water ($Ca(OH)_2$) milky (turbid)
Sulphides (S^{2-})	H_2S gas evolved with rotten egg odour, turns lead acetate paper black
Sulphites (SO_3^{2-})	SO_2 gas evolved with astringent odour, turns acidified $K_2Cr_2O_7$ paper green
Nitrites (NO_2^-) and nitrates (NO_3^-)	NO_2 brown fumes evolved with pungent odour, turns ferrous sulphate solution black. Brown ring test
Acetates (CH_3COO^-)	Colourless vapours with vinegar-like odour, turns blue litmus red
Acidified with sulphuric acid, 10% barium chloride solution	
Sulphates (SO_4^{2-})	Whitish turbidity forms $BaSO_4$, which is insoluble in acids but soluble in sulphuric acid only
Acidified with hydrochloric acid, 10% barium chloride solution	
Silicates	Whitish precipitation
Sulphates	Whitish precipitation

Tab. 23.4 (continued)

Reaction with	Observations
Thiosulphates	Yellowish sulphur-like precipitation
Acidified ammonium molybdate solution	
Phosphates (PO_4^{3-})	Yellowish turbidity
Silver nitrate solution	
Chlorides	White precipitate of silver chloride, soluble in ammonia but insoluble in nitric acid
Iodides	Yellowish precipitate of silver iodate, insoluble in ammonia, but soluble in sodium thiosulphate
Bromides	Pale yellow to cream precipitate of silver bromate, soluble in ammonia
Chromates	Reddish precipitate of silver chromate, soluble in dilute nitric acid
Hydroxides	White precipitate turns to brown silver hydroxide, soluble in dilute nitric acid
Carbonates	Yellowish precipitate of silver carbonate, soluble in dilute nitric acid
Thiocyanates	Insoluble in nitric acid
Starch-iodide paper	
Bromates	Turns purple
Chlorates	Discolours slowly
Chromates	Turns purple
Iodates	Turns purple
Nitrites	Turns purple
Dilute sulphuric + pot permanganate	
Arsenite	Discoloured
Bromide	Discoloured
Cyanide	Discoloured
Nitrite	Discoloured
Sulphide, sulphite	Discoloured
Thiocyanate	Discoloured
Thiosulphate	Discoloured

Tab. 23.5: Chemical spot tests for specific substances.

Substance sought	Common name of test	Reagents and procedure
Arsenic	Gutzeit method	Reaction of liberated arsine gas from acidified object with potassium iodide, to form a yellow stain on mercuric chloride paper
Gold	Acid test	Scratch the object onto a stoney surface, add a drop of nitric acid and if turns green, then copper is present not pure gold
Cyanide	Prussian blue test	
Plastic	Burn test	Melts and burns with a blackish smokey flame
Plastic Nylon	Solvent Burn test	Acetone Burns slowly with a celery-like odour
Nylon	Solvent	Hydrochloric acid or formic acid
Glass	Burn test	Melts, does not burn, yields yellowish colour to the flame. Only soluble in hydrofluoric acid
Wool	Burn test	Burns slowly with sputtering and self-extinguishers
Cotton, hemp, linen, ramie	Burn test	Catches alight readily, burns quickly, can burn with a yellow colour flame
Cotton	Solvent	75% sulphuric acid
Silk	Burn test	Melts and burns with a blackish smokey flame
Rayon	Burn test	Melts and burns slowly without ignition
Wool or silk	Solvent	5% sodium hypochlorite + 5% sodium hydroxide, boil to dissolve
Viscose	Solvent	60% sulphuric acid. Note most organics will react, sometimes forms darkish black colour (carbon)
Acetate	Solvent	Glacial acetic acid
Sugar, carbohydrates	Molisch's test	Add a drop of α-naphthol ethanolic solution to aqueous sample, then add sulphuric acid slowly down to the side of test tube and a purple colour is formed.

Data

Chapter 24
Data interpretation

Keywords: data, t-test, Student's test, Q-test, Grubbs' test, F-test, Gantt charts, control charts, regression analysis, calibration instrumentation, validation methods, errors, validation terminology, measurement errors, accuracy, precision, repeatability, reproducibility, Gaussian distribution, fishbone, minimisation errors, standard addition technique, least squares

This chapter covers the collection of data, analysis of data and the interpretation of laboratory test results and validation thereof using various statistical expressions and equations, as well as various aspects of record-keeping and reporting of results.

Some common basic terms used in the assessment of analytical data are as follows:

- **True result** is taken to signify the "correct" value for a measurement which remains unknown until a standard sample or a reference material is being analysed by the same procedure under identical conditions.
- **Accuracy** is the nearness of a measurement to the true value.
- **Error** is the difference between the true value and the measured value.
- **Mean** is the arithmetic average of a replicate series of test results.
- **Median** is the middle value of a replicate series of test results.
- **Degree of freedom** is an independent variable and is the number of degrees of freedom possessed by a replicate series of test results which **equals** the total number of test results in the series. The number of replicate results should be at least 25 measurements in order to be of any statistical significance.
- **Precision** is the variability of a measurement.
- **Spread** or **range** is the numerical difference between the highest and lowest test results in a series of replicates.
- **Coefficient of variation (relative standard deviation)** is often used for comparing the precision data from two sets of replicate test results.
- **Variance** is the square of the additive value of all the squared values of two or more sets of replicate test results. A useful figure when comparing the test results of a sample when tested by different analytical procedures exhibiting different levels of standard deviation.
- **Calibration curves** are compiled in order to determine the amount (concentration) of an analyte in a sample compared to a series of standards of known concentrations of the same substance as the analyte.
- **Measured** concentration levels:
  - If the concentration of the analyte is **too high** compared to the linear calibration curve (i.e. to the detection or workable range) of the measurement technique,

then a dilution can be made in a suitable pure solvent (usually lab-grade de-ionised water).

–   If the concentration of the analyte is **too low** for detection by an instrument's measuring capabilities, then the method of **standard addition** can be applied. In this procedure, a known pure quantity of the sought element or compound is added to the sample. The difference between the concentration added and the concentration measured is the original amount that was in the sample to be measured.

Data interpretation is one of the most critical aspects of analytical science.

There are many statistical techniques for the analysis of data. The field of **chemometrics** (analysis of raw data) is vast and forever expanding due to computerisation of advanced mathematical and statistical techniques used in the analytical sciences.

A very popular and powerful software programme is from IBM known as SPSS (Statistical Product and Service Solutions). It is basically a database and spreadsheet with many useful built-in equations and calculations, which is ideal for experimental research scientists and surveys.

The **reliability of test measurements** with its associative resources such as verification of resources, validation of methods and calibration of equipment must be utilised to its full capacity for any analytical work to be scientifically credible and acceptable to the scientific and legal communities.

The **limitations of the data** must be known so that unjustifiable conclusions are not drawn.

The laboratory analyst must evaluate the test result by comparing it with the actual result; if the result is out of the norm, then he/she should investigate whether any error has occurred during the analysis (or whether the sample itself was suspect). If the result is different from the norm, he/she must not change the result to fit the norm, as the result might be correct in showing a fault in the factory production, or a defect in the product.

If, after investigation by the laboratory analyst, the result is still suspect, but no errors can be determined, then another analyst from the same laboratory must repeat the test using freshly prepared reagents made by the second analyst. If similar test result is obtained, then the sample should be submitted to another laboratory for confirmation using the same or similar test methodology.

Note: Laboratory test results cannot be diligently compared between laboratories unless the same (identical) samples have been tested and the same methodology/test method has been used.

An estimate of the detection limit may also be made using one of the following:

–   take an instrument manufacturers' published detection limit and multiply this by a figure of 3, then the analyst would get a more realistic or workable detection limit for that particular analyte on that particular instrument; or

- take into consideration the concentration value that corresponds to the instrument's signal-to-noise ratio of 1:2, as per the manufacturer's specification for that instrument. Basically, this is twice the instrument's background noise readout.

24.1 The t-test: (aka Student's t-test)

This statistical test can be used to compare the averages of two sets of data, or one set of data against an accepted value.

Example:

Two analysts perform a series of titrations on the same solution. Are their average (mean) results (or standard deviations or variances) significantly the same or are they different (aka, as a two-tailed test)?

This statistical calculation or equation assesses or determines whether the mean of a set of data (e.g. measurements done by Analyst A) significantly differs from a specific value (called the hypothesised mean, aka, as a 1-tailed test), or from the mean of another set of data (e.g. measurements done by analyst B):

$$t - value = \frac{(A - B)}{s/\sqrt{n}}$$

where s is the standard deviation and n is the number of degrees of freedom (i.e. number of measurements).

If t-value is more than the one in the published table, then (it is not null) the two sets of data are significantly different.

If t-value is less than the one in the published table, then (hypothesis is null) the two sets of data are the same.

24.2 The Q-test (or Grubbs' test)

Example:

An analyst does 10 titrations on the same sample. Are all 10 results acceptable in order to calculate a mean and to use this mean figure in further statistical calculations (such as standard deviation) for the sample? Example: the $Q_{critical}$ value for n = 10 is 0.41 at 90% confidence level, from published tables.

The Q-test (Q = rejection quotient) is used to identify statistical outliers in a set of data. This test should never be applied more than once to the set of data. Arrange the set of data in ascending order; then from observation of the data, select the odd value out (outlier), Xa:

$$Q_{data} = \frac{|Xa - Xb|}{R}$$

where Xa is the suspect data point (e.g. a titration result that appears to be an out-lier), Xb is the next data point (i.e. result of another titration that appears to be clos-est to the outlier) and R is the range of the 10 data points (i.e. difference between highest result and lowest result, including the outlier).

The result, Q_{data}, must then be compared with a published table of critical Q values and read off the $Q_{critical}$ (at 90% or 95% confidence level) at n (number of measurements or degrees of freedom, e.g. 10):

If the result $Q_{data} > Q_{critical}$, then result Xa can be discarded (i.e. rejected) from the set of data.

If the result $Q_{data} < Q_{critical}$, then result Xa must remain in the set of data.

24.3 The F-test

Variance is the square of the standard deviation, σ. The ratio of two variances is known as the F function which is used to compare the precision of two sets of sepa-rate replicate data.

Example:

Compare the two test methods in order to decide if there is a significant differ-ence in the results from the different methods, or if they statistically give the same test result. The use of ratio of the coefficient of variances or the statistical deviations of the two lab methods (by convention, the larger variance is taken to be the numer-ator) is used; then compare the result to the published F tables for a certain n num-ber of determinations.

If F_{data} **exceeds** the critical value of $F_{critical}$ from published tables, then the dif-ference in variance or **precision** is deemed to be statistically significant (i.e. the dif-ference in precision of the two sets of data is not considered to be the same).

24.4 Gantt charts

A Gantt chart is a type of simple bar chart that illustrates a project schedule; e.g. start and finish dates of experiment and a summary of the elements of that experiment.

24.5 Control charts

A control chart is basically a type of graph used to illustrate how a process or test results changes over time, with respect to a set of standards. Data may be plotted in time order. The chart should always have a central line for the **specification value**, an upper line for the upper control limit and a lower line for the lower control limit. Other lines can be upper **warning limit** and upper **action limit**; these charts are known as **Shewhart charts**.

This information will exhibit whether a product or sample is in control or out of specification, e.g. it is within its tolerance limits. And check whether any action needs to be taken.

Various types of quality control (QC) chart formats are used by different laboratories that utilise different testing principles and techniques, such as a mean chart (accuracy), upper and lower range control limits like the precision or tolerance charts. Another QC type is the trend line charts. The spreadsheet Excel is very useful in plotting these trend lines.

A simple chart known as the **Cusum chart** (cumulative sum technique) is where a process is monitored on, say time interval periods and the data plotted on a chart or graph. Any deviations from a target result can be easily observed and hence quick action can be taken to avoid a drop or an increase in the values plotted.

Corrective actions and compliance audits must be taken by the laboratory management if there is any deviation from laid down controls by the standard operating procedures of university or corporation.

24.6 Regression analysis

It is a statistical process for estimating the relationships between variables or sets of data.

Linear regression uses the procedure of "**least squares**" to determine the best linear equation (e.g. straight line on an "'x" and "y" calibration curve).

$$\text{linear equation: } y = mx + c$$

where m is the slope of the line, curve or graph (this determines the sensitivity of the calibration (rate of change of x on y), c is the intercept of x with y where x is the vertical axis (e.g., optical density) and y is the horizontal axis (e.g., concentration of standard (calibration) solution). m and c are known as **regression coefficients**.

If the relationship between the measurement signal of the instrument and the amount of analyte is linear, the method of **least squares** may be used to obtain the **best straight line** through the data points; assuming that any amount of scatter or instability, observed in the measurements, is from the intrinsic variability of the instrument's signal. The instrument measurement signal is generally plotted on the y-axis and the amount or concentration of the standards and hence the sample analyte is plotted on the x-axis.

The best line is then calculated on the basis of minimising the variance (sum of the squares of the deviations (scatter, errors)) in y from the calculated line (i.e. the line (centroid) that is plotted through the centre of all the points). This line is then replotted as a best fit line from the regression analysis (least squares calculations).

The sample solution or analyte has to be read off or displayed by the graph or curve (this is mostly done today automatically by the electronics (microprocessor) of the instrument, but knowledge of the principal involved must be understood).

Many software packages that are included in the latest analytical instrumentation will automatically calculate and display the best fit for a calibration graph (curve). However, these can be misleading to the observer or analyst as they do not indicate (hide) where there is any scatter or noise on a data point.

A **good practice** when compiling a calibration curve is to have a series of calibration standards read twice on, say an atomic absorption spectrophotometer (AAS), is to have the standards measured before the samples are measured and then again after the samples are measured. The analyst can then observe if there is any **instrumental drift** during the measurements, by determining if the difference in readings of the standards, before and after, are significantly different.

24.7 Calibration of instrumentation

Note that many modern-day instruments can simultaneously separate, identify and quantify an analyte. Hence, it is critical that all analytical instrumentation is kept clean and in proper working order, by undertaking verifications and calibrations according to an organised time schedule. This should also include a schedule for preventative maintenance; example is the regular oiling of vacuum pumps or flow injection systems.

Calibration ensures that the instrument's measuring accuracy is compared to a known standard, usually an international reference standard.

Calibrations may be done internally or preferably outsourced to a special calibrating accredited laboratory.

Internally, it is done by following the instrument manufacturers' instructions on how to check the operation and accuracy of the detecting and measuring functions of the instrument.

Calibration and verification of how accurate and precise an instrument is operating, is undertaken by using various techniques, usually done by external accredited calibration laboratories (refer to ISO 17025).

Calibration certificates are issued by the outsourced calibrating laboratory, which is one of the mandatory requirements of accreditation of the testing or research laboratory.

Response factors and **calibration factors** must be checked regularly, at least every 2 years. Depending upon how frequently the instrument is used (e.g. daily), it might be necessary to re-calibrate more frequently (e.g. at 6 monthly intervals).

Verification of an instrument ensures that the instrument is correctly operated according to its stated manufacturers' operating manual and specifications; this can be done by running internal standards of known levels of concentrations during an analysis.

Instrumental errors are common in the analytical laboratory. Some of these can be summarised as follows:

- wavelength drift;
- wavelength shift;
- ageing of electronic circuit boards (e.g. dry solder joints or computer chip creep);
- misaligned mirrors and lenses of spectrophotometers due to transporting or moving sensitive equipment;
- mishandling of instrument accessories;
- dirty equipment;
- mass meters (balances) not levelled using the spirit bubble;
- unstable main electrical power supplies (e.g. voltage spikes and dips);
- using incorrect power cables with respect to wattage ratings;
- electrostatic interference, e.g. measuring millivoltage outputs from ISE and pH electrodes;
- broken or missing ferrite magnetic rings over power cables such as laptop cables, to reduce the electromagnetic interference; note that whenever there is electrical power (electrons) flowing through a cable, it induces a magnetic field around the cable;
- poor ground or earthing of sensitive electronic circuitry;
- contaminated electrodes, cells and cuvettes.

Calibrations are critical in determining and minimising these errors by applying correction factors to the instruments' outputs.

24.8 Validation of analytical test methods

What is validation? Is the test method suitable for its intended use?

Validation is a process of checking procedures, methods or test specifications, which have to fulfil the intended use, purpose or specific requirement of a process or test methodology.

Verification is to ensure that the process or instrumentation actually works; example: does the test method yield the analyte's presence and/or concentration in a sample?

Validation of processes, such as sampling, handling, transportation and storage of test items, is critical so that the test item itself, that is the sample, is received at the laboratory, with certainty as to its freedom from contamination, spoilage or degradation.

This may be referred to as the **stewardship** or **custodianship**, or even the **chain of custody** of the item (sample). Basically, it is the integrity of the sample as submitted for analysis.

Validation plan
- purpose of measurement;
- sample matrices;
- any chemical interfering substances expected;
- specific legislature or accreditation requirements;
- measurement scope;
- equipment and environment conditions;
- sample preparations and testing procedures;
- identification of performance characteristics;
- experimental design or ultimate plan.

Newly developed in-house methods, standard referenced methods and routine methods need to be validated to be effective in the laboratory, as per the mandatory requirements of international standards, such as ISO/IEC 17025 for general testing laboratories. There are similar standards such as medical and pathology laboratories.

Research laboratories should also follow these laboratory standards to add credibility to their published experimental and analytical research findings.

The techniques used for assurance and reliability of a test method and its validation may be as follows:
- The use of certified reference materials (**CRM**) to measure any bias of the method and its accuracy and precision.
- The systematic **assessment of any risk** or any external factors that might influence the test result; e.g. is there any bias present?
- The **robustness** or ruggedness of the method, that is, the tolerance to any variation of any measuring or process parameters, or external influences
- Intra-laboratory testing (**repeatability**) of numerous repeat tests of similar samples by in-house laboratory technicians
- Inter-laboratory comparisons (**reproducibility**) by other laboratories on the same samples (or similar sample matrix) using the same or very similar techniques or test methods
- Incorporating **QC samples** into the testing sequence
- Documented evidence of the method or analytical procedure that assures the **scientific veracity** and integrity of the test results; including regular checks for expired chemical stock reagents, preventative maintenance protocols and updating procedures
- Incorporating method performance characteristics such as measurement range **concentration linearity**, **sensitivity** and **selectivity**, limit of **detection (LOD)** and limit of **quantification (LOQ)**

In **summary**, the validation characteristics that must be researched, evaluated and documented for each test method are:
- accuracy of end result;
- precision of the end result;
- detection limit of test procedure and that of any instrumentation used;
- specificity of the analyte's ions or compounds;
- quantitation limit and linearity of the method measurement range;
- robustness to interferences, both chemical and physical;
- compliance testing of all variables of the analytical procedure;
- the calculated total value of the uncertainty of measurement (UoM).

The analytical data of the **test methodologies** (Fig.24.1), to be quantified or **validated**, are:
- **LOQ**: this is the lowest analyte concentration (i.e. the minimum quantifiable limit) that can be quantitatively detected with a stated accuracy and precision.
- **LOL** limit of **linearity**(dynamic range and working range): this is the direct proportional relationship between the concentration of the analyte and the quantitative measurement of that analyte. Hence, the dynamic range is that part of a straight-line relationship of quantitation and the test measurement.
- **LOD**: note that an instrument's stated or published detection limit (as specified in the instrument operator's manual) is NOT the same as the overall method detection limit (MDL), because other variables (such as dilution factors) have to be considered.
- **MDL**: it may be defined as the smallest amount (of analyte), i.e. the minimum concentration, that can be measured and reported within a stated confidence limit of 99%. The MDL for the complete analytical procedure will vary as a function of sample type and any sample pre-treatment procedures with subsequent solvent (usually aqueous) dilutions.
- **UoM**: there are many equations and calculations that are done to determine this important value which is a combination of all the characteristics of a methodology process.
- **Sensitivity** (S) or **robustness** of a method indicates the ability of the test procedure to determine a small amount of one substance (element or compound) in the **presence** of large amounts of **other substances**. This can also be stated as the **ratio of change in the response** of the method or instrument to that of concentration of the analyte being measured. Do not get confused with the LOD which does not incorporate sensitivity or selectivity which is dependent upon external interfering substances.
- **Specificity** is the effect that different sample matrices have on the final analyte test result. In other words, how specific is the measurement technique used to determine the concentration of an analyte in a sample?
- **Bias** is where there is a determined systematic difference between test results of a sample compared to that of a standard or QC control samples;

- **Recovery percentages** are calculated on the control or standard test results. Alternatively, it can be expressed as

$$\% \text{ Trueness} = \frac{\text{mean value} \times 100}{\text{added or expected amount}}$$

Note that a standard calibration curve or graph for the analysis of concentration of the analyte can show the levels of LOD, LOQ and LOL.

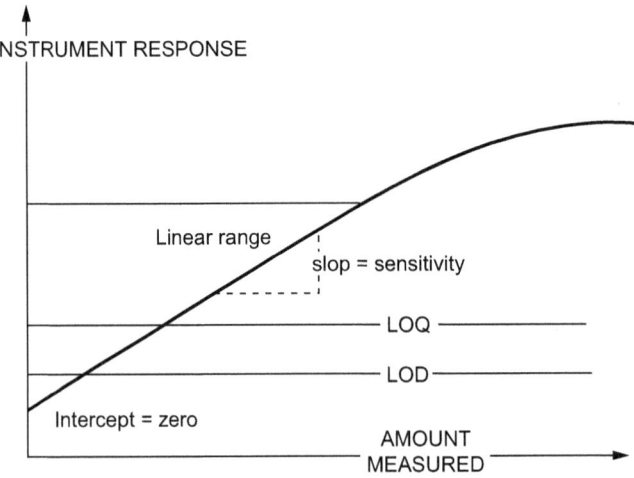

Fig. 24.1: Statistical terms.

Validation tools
- **Blanks:** reagent blanks are used to determine the effect of the analyte contributing to the test result compared to other factors (such as the acid or solvent used), which also contribute to the analyte's concentration test result.
- **Reference materials:** these are used to assess the accuracy of the analyte's test result.
- **Spiked samples or solutions:** these are used to assess the degree of recovery of the addition of a substance chemically the same as that of analyte's composition.
- **Repeatability:** this is replicate analysis to determine the precision of the analytical process.
- **Statistics:** determination of uncertainties of the measurements used in the analytical process.

24.9 Measurement of errors

What errors could be present?

Bias and random variations or errors are always present in collected laboratory data. Validations of techniques and methods (and calibrations of apparatus) are used for eliminating or reducing these errors and non-conformances of laboratory procedures.

Accuracy represents how close the test result is to the true value or an accepted reference value.

The true value is a theoretical correct value or a predetermined average result from a round-robin test programme on that particular type of sample. Also, an accredited reference sample or **CRM** of known value can be assumed to have a true value, and can be used as a control sample in the routine tests.

Precision is the degree of agreement between individual test results when a specific method is used repeatedly to more than one sampling (or harvesting) from a homogenous sample population.

Repeatability of a test procedure is the closeness of the agreement between the results of successive measurements of the same sample under the same conditions of measurement using the same apparatus, same analyst and same batches of chemical reagents.

Accuracy cannot be without precision, but precision can be without accuracy!

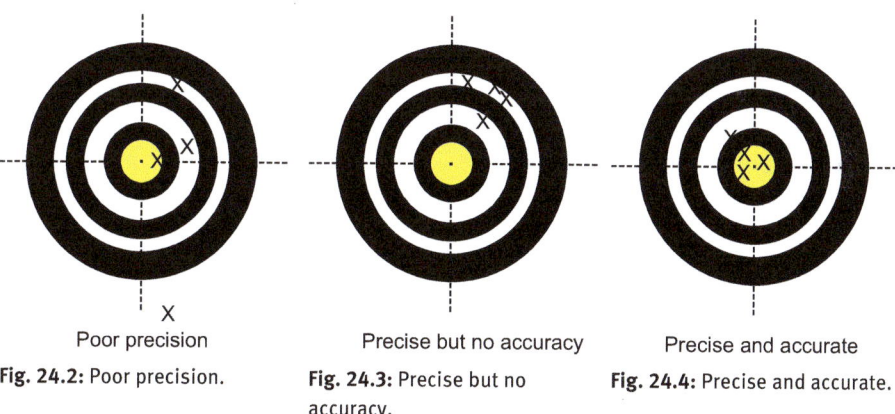

Poor precision

Fig. 24.2: Poor precision.

Precise but no accuracy

Fig. 24.3: Precise but no accuracy.

Precise and accurate

Fig. 24.4: Precise and accurate.

The test result cannot be more precise than the least precise measurement made of all those results used in calculating the final reported result.

Types of errors

There are basically only two important types of analytical errors:

What are determinate errors?

These errors are constant and systematic or proportional and can be measured, minimised and avoided, or correction factors applied.

Examples are:
- incorrect sampling, or insufficient sample for the analysis;
- errors of test method, such as incompleteness of a reaction;
- equipment errors, such as pH meter out of calibration;
- impurities in reagents;
- expired chemical reagents or standardised solutions;
- contamination of acids and/or alkalis;
- personal errors, such as colour blindness of laboratory worker when performing titrations;
- operational errors, such as overfilling or underfilling volumetric flasks, or filling volumetric flasks to mark at incorrect temperatures;
- proportional or additive errors, such as errors in weighing;
- unskilled laboratory workers.

What are indeterminant errors?

These errors are accidental or random and cannot be measured or assessed, i.e. unknown errors due to causes which the analytical chemist has no control, but are governed by the laws of probability (**Gaussian distribution** curve) of frequency of error occurrence. This Gaussian curve illustrated below in Fig. 24.5; is also known as the normal probability curve. Basically, the errors are:
- small magnitude errors occur frequently;
- large magnitude errors occur seldom;
- positive and negative errors occur with equal frequency and with equal magnitudes.

The analytical scientist must also be aware that there will be an occasion when a single random (test) result will be completely out of the probability curve, without any scientific explanation at all; nature has a way of throwing us a curve ball from time to time.

The **Gaussian** (bell-shaped curve) curveillustrated below in Fig. 24.5, in analytical chemistry, represents the degree of precision of a test result, namely the standard deviation. This curve is also known as the normal error curve. The ordinate scale (y-axis) represents the frequency of a certain event occurring; in other words, the number of times a test has been repeated on the same sample by the same analyst (in order to calculate the standard deviation of that test). The abscissa (x-axis) represents what that event is (e.g. a test for fat in petfood); in other words, the actual results of the testing procedure.

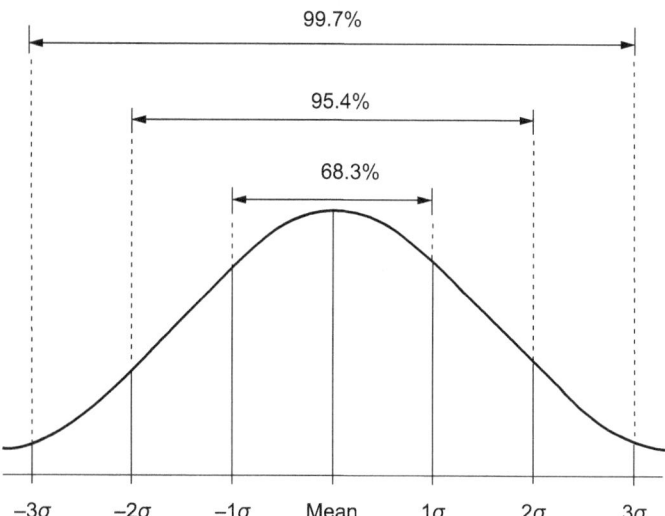

Fig. 24.5: Gaussian curve.

In Gaussian curve, σ is the standard deviation from the mean and the mean is considered to be the true value of result from a laboratory test.

This means that a standard deviation of 1σ indicates that the test result will have a confidence level of 68.3%.

And at 95.4% confidence level of 2σ indicates, in general, an acceptable target of a test result.

What all this means is that if a sample or test item were to be tested 100 times, by the same method and analyst, at a stated 95.4% confidence, then the test result would yield correct 95 times and would have 5 incorrect results. The final result would be reported as having $\pm2\sigma$.

Practical example of determining fat content in a petfood gives the following information:

Standard deviation (2σ) of the method was ±0.35 g/kg.

Test result of the fat sample was 12.55 g/kg.

Hence, the analytical chemist would report the result of the test as:

Fat content of sample = 12.55 g/kg ±0.35 g/kg, at the 95% confidence level.

Absolute error of a result is the difference between the observed or measured value and the true or most probable value of the quantity measurement.

The amount of this error could be either positive (+) or negative (−), i.e. the test result could be either too high or too low!

Relative error is the absolute error divided by the true or most probable value and is measured usually as a percentage.

Repeatability is the calculated standard deviation of a series of test results undertaken on the same sample simultaneously by the same analyst in the same laboratory. This figure basically depends upon the skill of the analyst.

Reproducibility is how closely another laboratory can repeat the analysis on the same or similar sample, using the same or similar test methodology. This figure would also indicate the scientific correctness of the test methodology used for the analysis by both laboratories.

On combining the concepts of repeatability and reproducibility, another concept is calculated known as the **gage capability**.

A common **root cause analysis** (RCA) technique (a cause and effect chart) used for quantifying tests errors in analytical chemistry is the **fishbone diagram** see Fig. 24.6 below as an example; (also known as the Ishikawa diagram, named after its inventor):

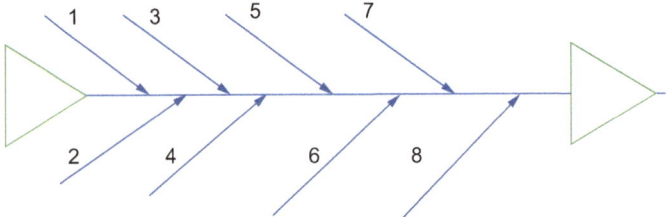

Fig. 24.6: Fishbone diagram used for RCA.

Example: a titration of aqueous diluted alkaline sample with standardised 0.1 N acid:
1. Mass weighed out of sample was 2.3456 g (tolerance of mass meter = ±0.0001 g)
2. Weighed sample was dissolved in water and made up to 100 mL volume (tolerance of volumetric flask = ±0.5 mL)
3. Volumetric flask calibrated at factory to 20 °C but sample solution is at 25 °C (temperature error of glass flask causes volume error of ±2 mL)
4. A dilution of 10 mL in 100 mL was undertaken (tolerance 10 mL pipette ±0.1 mL); analyst's pipetting error causes a volume loss of 0.2 mL
5. The dilution for second 100 mL flask was ±0.5 mL; and volumetric flask calibrated at factory to 20 °C but sample solution is at 25 °C (temperature error of glass flask causes volume error of 2 mL)
6. 50 mL burette used had a tolerance of ±1 mL, including error of analyst reading meniscus ±0.2 mL
7. Titration end point of titre of 34.9 mL error of ±0.2 mL
8. Standardisation normality error of the 0.1000 N acid titrant was 0.0998 N (±0.0002 units)

The equation is

$$\frac{34.9 \text{ mL} \times 10 \times \text{factor (which is a constant)} \times 100}{2.3456 \text{ g sample} \times 100} = \%\text{FFA of sample} \pm$$

Minimisation of errors

Determinate errors can be reduced by the following procedures:

- Calibration of equipment: all apparatus and instrumentation must be calibrated by certified calibration laboratories, in order to determine errors such as any glassware volume error or instrumental operational faults.
- Blank determinations: here, testing is done simultaneously with the sample, but in another determination, testing is done without the sample. Basically, testing is done for both determinations under the same conditions, reagents and so on.
- Control determination or internal standard procedure: here another determination is undertaken but where the sample is substituted with a standard substance of the same constituents or matrix as the sample. This is normally referred to QC controls.
- Duplicate or triplicate determinations or undertaking parallel testing on the same sample.
- Different sample weights of the duplicates are another procedure and then to compare the final test results to determine the error difference. This difference multiplied by a factor of 3 could be used as a quick indication of the degree of UoM.
- Independent analysis is where the same sample is tested by two or more different techniques or methods such as by gravimetric and volumetric analysis.
- Standard addition.

The laboratory analyst must also undertake the following steps to minimise errors and to add credibility to his/her test results:

- Correctly calibrate instrument and apparatus, or apply any calibration factors to measurements.
- Work in a clean and tidy environment, to avoid contamination of the sample, glassware and so on.
- Run a reagent blank determination with the sample test (to compensate for any impurities in reagents used); i.e. run simultaneously another "sample" but without the sample.
- Run a control sample (a previously tested sample) or a standard (a certified test piece) simultaneously with the sample test (to compensate for operational or instrumental errors).

The amount of the constituent (analyte) in the sample can then be corrected for errors by calculation from the equation:

test result found for standard or control = stated amount of constituent in standard or control;

test result found for sample = amount in sample to be reported.

– Use another method to check accuracy of result (this checks the accuracy of the methodology used); the relative amount of constituent in the sample determines the choice of test method to be used.
– Analyse sample in duplicate or triplicate (to establish precision of the laboratory analyst's skills and to detect any determinate errors but does not detect any "constant" errors); also by varying the amount of sample tested (e.g. take 5 g sample and also take 10 g sample, test simultaneously) can indicate the presence of any errors in the methodology or analyst's skill.
– Use the method of **standard additions** (to determine whether any accumulative errors present). This method (aka laboratory-fortified matrix) is also ideal for determining trace quantities of an analyte/constituent in a complex matrix sample; also, if % recovery of a "spiked" sample is greater than 95%, then this establishes confidence in the accuracy of the test method.
– A common procedure in organic analyses is the internal standard, which is used generally in HPLC and GC work. This is where another substance of a definite concentration, similar to but not identical to, the analyte, is added to the sample and tested.
– Purification of reagents by re-crystallisation.
– Minimise the signal-to-noise ratio (S/N) by optimising the operation of the instrument being used; example is aligning the hollow cathode source lamp in AAS measurements.
– Environmental noise, such as electrical voltage power spikes or dips, in the mains supply. Also, nearby cell phones' radiation interference (and electrostatic charges) on sensitive analytical measuring instrumentation.

UoM
A very critical figure which gives credibility to the analytical chemist's analytical testing abilities and skills. This figure must be stated in final laboratory reports and data spreadsheets.

There are several reagent algorithms for calculating this value from a series of test result inputs.

Statement of conformity
This is where a document specifically states that something (object, item or sample) fulfils certain mandatory (chemical, physical or otherwise) requirements with respect to quality or standard specifications.

Principle of **standard addition**

Basically, the principle is comparing test results to those of a series of a measured quantity of standard that has been added to the sample. This technique is generally used to detect and measure analytes that are below the detection limits of instrumental analysis but is a powerful technique to also detect measurement errors. This is similar to calculating the percentage recovery or a test method.

The accuracy and precision advantage here are that the standard and unknown (analyte) are measured under essentially identical conditions.

Properties of chemical (primary) standards or compounds (CRM) that are used to calibrate, validate or check the accuracy of instrumental or chemical measurements of an analyte, are as follows:

– Stability properties such as not being deliquescent, efflorescent or hygroscopic
– Temperature and humidity stable
– In an easy form to handle under safe conditions
– Preferably stable in an aqueous solution
– High purity (above 98.5%) with a certificate of analysis from the manufacturer, including the mandatory SDS document
– Readily available and reasonably low cost, from laboratory suppliers with an expiry date stamp
– It should not be toxic or radioactive and must be safe to handle without any after-effects to the laboratory analyst.

Chapter 25
Laboratory reports

Keywords: rules of significant figures, lab report

In this last chapter, the most important aspect of all tasks that an analytical chemist is faced with is the final laboratory report.

Common practice dictates that an experimental or final test result should be rounded off to the correct number of digits, so that it contains **only the digits known with certainty,** in addition to the first following uncertain digit.

25.1 What are significant figures?

The **rule of significant figures**: all digits that are certain, plus one (the last digit) which contains some uncertainty, are said to be significant figures. Thus, they are the digits of a number which are needed to express the precision of a measurement from which the number was derived, e.g. if (surplus) digit to be rejected is a 5 or above (i.e. when rounding off a 5), then add one to the last significant figure; i.e. when rounding a 5 always rounds off to the nearest **even number**.

For example, 61.455 becomes 61.46, or 61.465 becomes 61.46 (four significant figures).

In general, when test results are reported, it is assumed that the last digit is always ±1; e.g. a result reported as 3.68% generally means that this result is 3.68% ± 0.01% (three significant figures, or to the nearest two decimal places).

But, if the standard deviation, σ, is known for that test procedure, say ±2%, then the result will be reported as 3.68% ±2%, which equates to a result of **uncertainty** between 3.75% and 3.61%.

All digits from measurements or digital displays should be included in the final calculation. The final result should then be corrected to the **least number of significant figures** that were in the respective analytical measurements and this also applies when reporting the final result to the **least number of decimal places** in the figures used in the final calculation.

In the practical analytical world of notation, no more than **four significant figures** should be reported as a final test result.

In microbiology and other sciences, the expression of, for example, the final bacteria count or other measurements could be recorded differently, such as a total bacteria count of 26.5×10^{-3} colony forming units (where the actual average count was 265 bacteria on a series dilution of three repetitions).

https://doi.org/10.1515/9783110721201-026

Examples:
- 0.0005 has 1 significant figure (i.e. result could be 0.0004 or 0.0006)
- 498 has 3 significant figures (i.e. result could be 497 or 499)
- 49.80 has 4 significant figures (i.e. result could be 49.79 or 49.81)
- 4.979×10^5 has 4 significant figures (i.e. result could be 4.980×10^5 or 4.981×10^5).

Rules
- The digit zero is a significant figure except when it is the first figure in a number or when used in powers of ten.
- When numbers are added or subtracted, the number that has the fewest decimal places (not the fewest significant figures) determines the justifiable number of decimal places to the answer:

$$e.g.\ 0.012 + 0.0126 + 0.14 + 105.3 = 105.4646 = 105.5$$

Reported result becomes 105.5 as the acceptable answer to four significant figures, although the terms 0.14 and 0.012 have only two significant figures.
Note that the rounding off must be done *after* the addition or subtraction.
- In multiplication or division, the answer cannot contain no more significant figures than the least accurate measurement.

The % precision of a product or quotient cannot be greater than the % precision of the least precise factor entering into the calculation.
Note that the rounding off must be done **after** the multiplication or division.
Example (where standard deviation values are given):
An analytical result is reported as 1,857 mg/L and the analyst is certain of the correctness of the 185 digits, but is uncertain as to whether the last digit 7 should be a 6 or an 8, because of unavoidable uncertainty in the analytical procedure or test method used.
In general, the last digit in any reported number is uncertain by ±1.
However:
- if the **standard deviation was known** for this analytical procedure from previous experimental work to be **±10** mg/L, then the analyst would have rounded off the result to 1,860 mg/L before reporting it as 1,860 ± 10 mg/L (i.e. three significant figures because 6 is uncertain); or,
- if the **standard deviation was known** for this analytical procedure from previous work to be **±100** mg/L, then the analyst should have rounded off the result to 1,900 mg/L, and report it as 1,900 ± 100 mg/L (i.e. two significant figures).

Thus, report only such figures as are justified by the accuracy and precision of the test methodology.

25.2 The report

The technical report (or laboratory report) is one of the most important and critical means of reporting or submitting laboratory test information and data without any bias and with minimum experimental errors.

The report must be written in "good" English without any possibility of misinterpretation by the reader or reviewer. "Good" English is the use of writing with all necessary punctuation, past tense and no personal pronouns. Remember that the analytical chemist is not writing a novel or fiction (i.e. formal and rather impersonal language should be used).

The past tense is a **grammatical tense** that places an action (verbs) or situation in the past of the current moment.

Examples of **personal pronouns** are:

I, we, me, my, us.

An example of **critical punctuation** is:

"let's eat Grandpa!"

"let's eat, Grandpa!"

Both sentences contain the same words in the same sequence but have completely different meanings.

The type of text used should be the Arial or Times New Roman font family, with a font size of 12 pixels, font style regular. Page parameters are:

Margins:

Top 2.54 cm; bottom 2.54 cm; left 1.91 cm; right 1.91 cm.

Paper size:

A4 or letter.

Paragraph:

Line spacing: 1.5.

Alignment: justified.

A basic format for a technical or scientific report could be as follows:

The sequence of the front pages should be numbered with lower-case Roman numerals, using the hierarchical numbering system:

- Report or project **reference number**: that is, the analyst's tracking or job number.
- **Title** of report: written in bold upper-case letters.
- Name of **author**: senior investigator's name and those of any co-authors, if more than two co-authors, then use the name of principal author followed by the Latin word "et al."
- **Abstract** or summary: an indented short paragraph, in italics, of maximum 15 words of the objective, how it was achieved and what conclusion was made; text must contain keywords in bold.
- **Keywords**: a list of the most important words or phrases that have been used in the main text of the report or thesis.

- **Table of contents**: only if the report is very lengthy, e.g. longer than seven pages.
- **Acknowledgements**: basically a "thank-you note" to any supervisor, professor or mentor who has assisted with the report. This section may be placed in the preface or bottom page notes, reference or bibliography.

The sequence of headings and subheadings should be numbered with Arabic numerals, as below, using the hierarchical numbering system:

1. **Introduction**: an outline of the plan of action, i.e. what the report is about and its context. This should not be more than a single paragraph.
2. **Objectives**: what the work being reported was expected to achieve, similar to a hypothesis (a tentative explanation for an observation, phenomenon or scientific problem that can be tested by further investigation).
3. **Theory**: if applicable to the experiment, project, lab test or report. Example is any background theory needed for the reader to understand the report.
4. **Method**: a test procedure or methodology, which would include where applicable, subheadings for apparatus or equipment used and/or (chemical) reagents used.
5. **Results:** only the critical data (or measurements) obtained directly from the experiment, project or laboratory test results. Preferably tabulated (e.g. Excel spreadsheet) for easy reading and understanding, followed (where necessary) by graphs, diagrams, printouts and spreadsheets; even video or PowerPoint.
6. **Discussion or interpretation**: how the results (5 above) correlate with your objectives (2 above); may also include equations, calculations and any variables that might affect the objective, etc.; otherwise these may be listed in Appendices.
7. **Conclusions**: the analyst's opinion or statement of what has been achieved, whether positive or negative, with respect to the introduction (1 above) and/or objectives (2 above).
8. **Recommendations**: if any, based on conclusions/opinion (7 above).
9. **References** and/or bibliography: an acknowledgement to any literature source that has been consulted. The citation must be sufficient to allow the reader to look up, or link to, the reference.
10. **Signature:**the report should contain the signature of each senior analyst or chief researcher who has rendered an input of any significance or critical aspect of the work undertaken, and report must be dated.

Be very careful that no **plagiarism** was used (the deliberate or reckless representation (copying) of another person's words, phrases or ideas as one's own without acknowledging that person's work).

There are many different ways of reporting the order in which the elements of the **reference** should be stated.

Example
"The Effect of Nitrates on Platinum laboratory ware at temperatures above 1,000 °C", Joe Blogg, Internal RSA Communication, Vol II, 3013.

Appendices: here place the graphs or charts, diagrams, spreadsheets, tables, equations, calculations, a list of abbreviations or other appropriate information pertinent to the report, conclusion, hypothesis or opinion. The appendices should be listed in Roman numerals.

Glossary: a list of uncommon technical or scientific terms that has been used, as well as any pertinent words or phrases.

NOTE
Correct sentence construction is critical in any scientific or technical document!
What we write, type, SMS texting or tweets and submit to others gives them an impression of how educated and professional we are in communication to each other.
Here are some pointers:
- Do not end sentences with a preposition.
- Practise on how and when to use conjunctions, nouns, adjectives, adverbs and verbs.
- Write short sentences, know when and how to use phrases and clauses.
- Keep the same tense throughout the document (e.g. for reports, write in the past tense; for standard operating procedures, write in the future tense; for test methodology, write in the present tense);
- Avoid writing ambiguous sentences or instructions.
- Use correct punctuation wherever necessary.

The **International System of Units** (SI) should be used in all scientific thesis and technical reports.
For laboratory test results, the following unit concentration formats should be observed:
- If result is less than 1 g/L, then result should be reported in mg/L, i.e. milligrams per litre.
- If result is less than 1 mg/L, then result should be reported in µg/L, i.e. micrograms per litre.
- If result is less than 1 µg/L, then result should be reported in ng/L, i.e. nanograms per litre.
- If result is greater than 10,000 mg/L (i.e. 1.0%), then result should be reported in % m/v, i.e. grams per 100 mL.
- If result is greater than 10,000 mg/kg, then result should be reported in % m/m, i.e. grams per 100 g.

Note the rules of significant figure and decimal points.

It is important to note that the result of any analytical procedure can **never be better than the correct representative sampling procedure** and the correct diligent sample preparation on which the analysis was undertaken.

Refer the Standard Practice for Reporting Opinions of Scientific or Technical Experts (ASTM E620), for a comprehensive guide on how to express an expert opinion in a written report. Other useful guides are ASTM E678 and ASTM E860.

Appendices

Appendix I
Laboratory safety and checklists

A list of *all* chemicals used and stored should be kept; also, a material safety data sheet, now known as a **safety data sheet** (SDS), for each chemical must be kept and easily available in the laboratories. The SDS lists the chemical's physical, chemical and hazardous properties, as well as the chemical's various regulatory listed numbers.

Flammable agents are Class 3 (liquids) and Class 4 (solids).

Oxidising agents and organic peroxides are listed as Class 5.

Poisons are Class 6.

Corrosive substances are Class 8.

The chemist must always check the bottle label and hazard pictogram to see which class the chemical is listed. Also check the expiry date printed on the package.
 Laboratory technicians and workers must note the following:
- **Flash point** is the minimum temperature at which the vapour pressure of a liquid is sufficient to form an ignitable flash (explosive) mixture with air near the surface of the liquid. Once ignited, it becomes **self-extinguishable**.
This test is usually performed in a **standardised** piece of apparatus with an **open cup** or **closed cup** operation.
- **Fire point** is the minimum temperature at which the vapour or gas mixture of a liquid with air is sufficient to maintain a flame or **continue to burn** near the surface of the liquid. This temperature is about 5 °C above its flash point.

https://doi.org/10.1515/9783110721201-027

- **Ignition point** or **autoignition temperature** is the minimum temperature necessary for a self-sustained combustion (fire) of a substance in the **absence** of any external source of ignition (e.g. flame, spark). This temperature is generally above 200 °C for most materials.
- **Upper and lower explosive limits** (refer to Fig. AI.1 below) or upper and lower flammable ranges are the **range of concentration** (by volume) of the chemical **vapour in air** for which a flame (i.e. combustion; reaction with oxygen) can take place or propagate. If the mixture is too rich with the vapour (i.e. it exceeds the upper limit), or if insufficient air (oxygen) is available, then combustion (fire) cannot take place. If the mixture is too lean (i.e. below the lower limit), then combustion (fire) cannot take place. A fire or explosion cannot occur unless all three sides of the Fire Triangle are present; namely air, combustible matter and source of ignition; refer to Fig. AI.2 below.

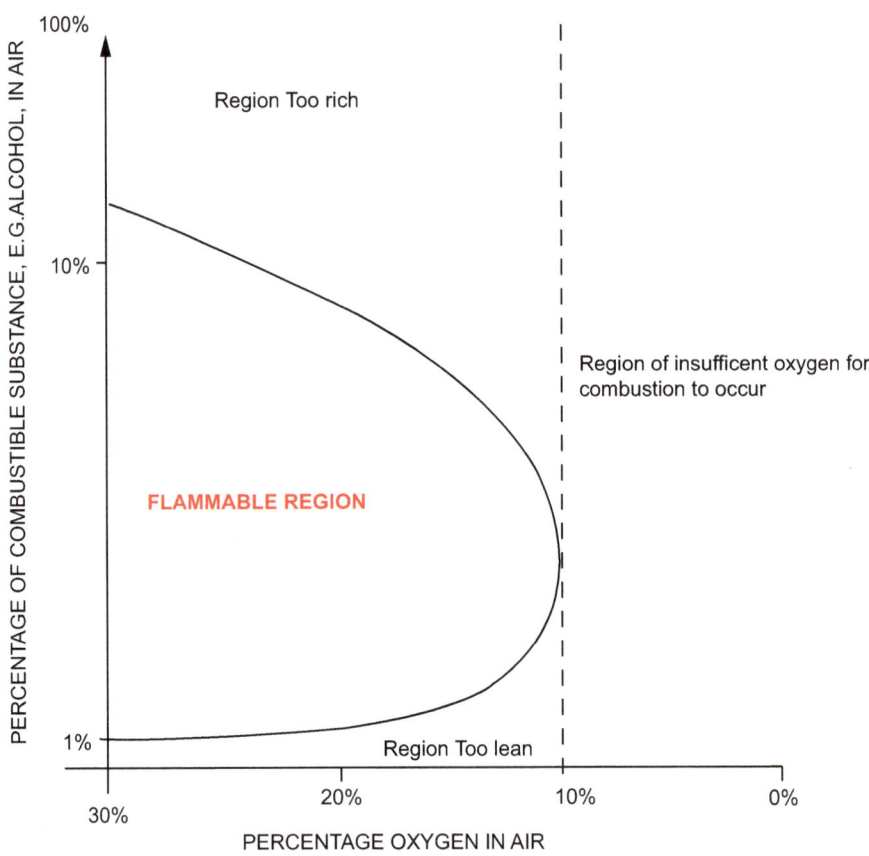

Fig. A.I.1: Upper and lower explosive regions.

FIRE TRIANGLE

NOTE:

1 Liquids and solids DO NOT BURN, it is the vapours
that are given off that burn.

2 Also the SOLID must be in the correct form, e.g. steel-
wool not steel; or wood shavings.

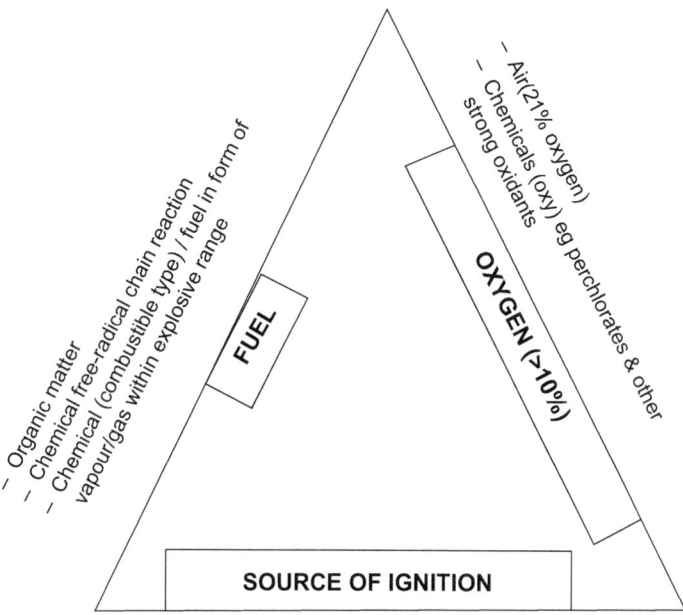

– Chemical reaction: exothermic (gives off heat)

pyrophoric (eg magnesium metal)

welding/cutting

– Electrical: static

sparks

heat/hot surfaces

– Mechanical: friction: heat

sparks

heat (compression of gas)

– Spontaneous combustion (eg oily rags) (heat)

– Auto/spontaneous-ignition (presence of high temperatures)

– Open/naked flame or smoking

– Sunlight through glass (magnification)

– Arson

Fig. A.I.2: Fire triangle.

- **Pyrophoric chemicals** are those chemicals or mixtures that will **ignite spontaneously** (burn or explode) when **exposed to air**, e.g. iron sulphide.
- **Water-reactive chemicals** are those chemicals or mixtures that will **react violently** when **exposed to water** or aqueous mixtures, e.g. sodium metal.
- **Shock-sensitive chemicals** are those chemicals or mixtures that will **react violently** when exposed to a **mechanical shock or impact**, e.g. a motor vehicle's safety airbag. Picric acid is such a common chemical used in analytical chemistry; it is dangerous.

A **safety shower** is a must in any chemical laboratory that handles toxic or flammable chemicals, acids and strong alkalis.

A **fire blanket** is a must in laboratories that handle flammable solvents or strong oxidisers.

Peroxide formation can be very dangerous when the formation is unintentional or unexpected, e.g. ethers in enclosed stock glass bottles, or **Winchester** amber glass 2.5 L bottles.

It is advisable to store large quantities (200 L or more) of flammable chemicals in a specially built registered chemical **storeroom**, with fireproofing and adequate drainage, as per local city building bye-laws and insurance requirements.

Other examples of extreme **hazardous chemicals** are:
- **Bromine liquid**, saturated bromine water, is used in oxidation reactions, for converting sulphites to sulphates; the fumes are extremely corrosive
- **Sodium metal** is stored under mineral oil or paraffin (kerosene); explosive (hydrogen evolved) with water.
- **Hydrofluoric acid** burns (these are normally not visible immediately on contact with skin!).
- **Potassium metal** is stored under mineral oil or paraffin (kerosene), explosive with water and highly reactive in the presence of air (oxygen).
- **Phosphorous** metal (stored under water).

These should be isolated and stored separately from each other.

It is unsafe to store certain chemicals in close proximity with one another, e.g. never store oxidising chemicals, such as chlorates, nitrates and perchlorates, with reducing chemicals or with organic materials.

Hazardous combinations of chemicals must be avoided to reduce the risk of fire, explosion and/or evolution of toxic fumes.

Never store any oxidising chemical next to any reducing chemical (Tab.AI.1).
Some general **laboratory rules** are:
- The playing of games or other similar activities in a laboratory is prohibited.
- Protective clothing and other safety equipment (PPE) must be used at all times.
- No unauthorised persons to enter the laboratory.

Tab. A.I.1: Hazardous chemicals.

Oxidising chemicals	Reducing chemicals
Chlorates	Ammonia, ammonium hydroxide
Chromates	Carbon
Chromium trioxide	Metals
Dichromates	Metal hydrides
Halogens	Nitrites
Hydrogen peroxide	Organic compounds
Nitric acid	Phosphorus
Nitrates	Silicon
Perchlorates	Sulphur
Peroxides	–
Permanganates	–
Persulphates	–

Tab. A.I.2: Flame and heat temperatures (approximate).

Source	Temperature (°C)
Fluorescent light	70
Incandescent light	200
Cigarettes	5
Hotplate	450
Match	700
Candle flame	800
Electrical spark	1,000
Bunsen burner (depending on gas mixture type)	400–1,300
Wood/charcoal	1,400
Hydrogen/air	2,000
Carbon monoxide/air	2,300
Acetylene/air	2,600
Acetylene/oxygen	3,300
Electrical arcing	3,500

Tab. A.I.3: Explosive limits, auto and flash points (approximate).

Gas or vapour	LEL (% v/v)	UEL (% v/v)	Auto-ignition point (°C)	Flash point (°C)
Acetaldehyde	4	57	175	−30
Acetic anhydride	3	10	310	50
Acetone	3	13	460	−18
Acetylene	3	80	330	−18
Acetic acid glacial	4	20	470	40
Ammonia (gas)	16	25	630	11
Aniline	1	11	600	76
Benzene	1	7	560	−11
n-Butane	2	8	400	−60
Butyl alcohols	1	11	340	11 to 28
Carbon bisulfide	1	44	90	−30
Carbon monoxide	13	74	600	−191
Cyclohexane	1	8	240	−17
Dioxan	2	22	180	18
Ethane	3	13	480	135
Ethyl acetate	2	12	430	−4
Ether	2	48	150	−29
Ethyl alcohol	4	19	360	11 to 13
Ethylene	3	42	490	1
n-Heptane	1	6	200	−4
n-Hexane	1	7	200	−22
Hydrogen	4	74	400	0
Hydrogen sulphide	4	46	270	1
Methane	5	14	580	1
Methyl acetate	3	16	450	−13
Methyl alcohol	7	36	380	11
Methyl ethyl ketone (MEK)	2	10	400	0
n-Pentane	2	8	300	−40

Tab. A.I.3 (continued)

Gas or vapour	LEL (% v/v)	UEL (% v/v)	Auto-ignition point (°C)	Flash point (°C)
Petrol (gasoline)	1	8	280	−43
Petroleum ether 40/60 °C	1	6	280	−46
Propane	2	9	460	−104
Propyl alcohols	2	12	400	11 to 15
Pyridine	2	12	480	20
Toluene	1	7	480	4
Xylene	1	6	460	22

- Know your institute's emergency procedures and exits.
- Know where the fire-fighting equipment and extinguishers are kept.
- Know your institute's safety policy and disciplinary code.
- Never leave an open flame (Bunsen burner, spirit lamp or an AAS flame) unattended.
- Practice good housekeeping at all times; there is a place for everything and everything in its place.
- Know the properties of the chemical that is used in experiments or analyses, namely, its:
  flammability, reactivity, corrosiveness, toxicity and stability; refer to its SDS, or the R (risk) and S (safety) phrases on the bottle label.
- Remove any spillages immediately using the correct procedures; refer to its SDS or the R and S phrases on the bottle label.
- Discard all broken, chipped and badly scratched glassware into a separate waste container, labelled Broken Glass.
- Discard all reagents/solutions that are not in labelled bottles or are "out of date"; that is, expired.
- Wash hands frequently to remove any traces of chemicals.
- Eating, drinking, vapouring or smoking should not be permitted in chemical laboratories (except where tasting is done in food laboratories where organoleptic examinations are done under controlled conditions).
- **Always wear certified safety glasses or goggles** whilst working in the laboratory!
- **If a gas fire occurs, then immediately turn off the supply of gas**, e.g. at valve of gas cylinder.
- **If a solvent fire occurs, then allow it to safely burn itself out, making sure that there are no other combustibles in the vicinity,** because a fire extinguisher would just blow the burning liquids in all directions!

- The **fire triangle** below is important to remember. If any one side of the triangle is removed, then a fire (or explosion) cannot occur. If all three sides of the triangle are present in any situation, then a fire (or explosion) *will* occur!

Some safety hints
- Never leave vessels **unattended** when evaporation of liquids is in progress; the vessel (glass, porcelain) may crack or explode as "dryness" condition is approached (e.g. presence of perchloric acid in sample digestions) if the heat source is not adjusted correctly; lower the temperature gradually as the liquid level decreases.
- Always use caution when removing glassware from a heat source and avoid placing on a cold or damp surface; **sudden temperature changes** may cause the glassware to **explode** or crack.
- Never place a red-hot item from the furnace directly into a desiccator for cooling, as the thick glass of the desiccator or a glass cabinet will violently expand and crack; cool the item temporarily on a heat resistance mat or top of the furnace and then place into desiccator.
- Always **heat or cool vessels slowly** to prevent thermal shocks to the vessel.
- Always diffuse the heat source (Bunsen burner) by using a metal gauze, or air/water bath or sand bath.
- Always use **anti-bumping** devices in the vessel (such as powdered pumice and glass beads) when rapid heating of the vessel and its contents are required.
- When using glass or metal mechanical stirrers in a glass vessel, always predetermine the height of the stirrer before use to ensure there is no contact between the stirrer blade and the bottom or sides of the vessel. For magnetic stirrers, regularly check the speed (rpm) to prevent splashes.
- Never try to pull a glass thermometer out of a rubber bung, rather carefully cut away the bung (wear gloves or some other sort of hand protection).
- The use of water or glycerol (glycerine) is recommended on both tubing and rubber bungs when inserting glass tubing into a bung (wear gloves or some other sort of hand protection).

Properties of oxygen
Normal concentration in air = 20.8%
Deficient for short periods or lack of air = 16%
Minimum for life support = 18%
Prevention of combustion of materials = 5% or less

Hazardous properties of ammonia gas
Explosive limits; LEL = 16%; UEL = 25%
Auto-ignition temperature = 650 °C

Tab. A.I.4: Oxygen levels.

Environment	% Oxygen level in air
Dry air	20.8
Exhaust breathe of human	18
Safe to breathe	18.5–23.5
Unsafe for humans, anoxia begins	<16
Fire hazard, enhanced combustion	>24
Fire/oxidation to occur (minimum to support combustion)	>10
Prevention of fire or explosion	<5
Inert atmosphere in enclosed container	<5

Safety inspections

These are performed regularly by a nominated and elected safety representative (SHE is safety, health and environment) in accordance with the mandatory legal requirements of the Occupational Health and Safety Act.

There are many laboratory safety checklists on the Internet. Choose one that is suitable to the type of laboratory (e.g. pathology lab).

Fire extinguishing equipment checklist

(a visual checklist)

Name of competent person or safety representative: ....

Name of responsible person or management: ....

Date of inspection: ....

Signatures of:

Check item:	Yes	No	Action by whom:	Expected date of completion:
Is equipment labelled and in its assigned position, and weather protected if necessary?				
Are the extinguishers of the correct type for the specific fire hazards?				
Are the operating instructions on the equipment visible from all sides?				
Are the extinguishers accessible and clearly visible with appropriate signage?				
Is the equipment free of corrosion and physical damage?				
Are the extinguishers free from any signs of leakage?				
Are the nozzles free from blockage?				
Are the pressure gauges in working condition? (tap lightly on gauge)				
Are any extinguishers due for reconditioning, servicing or refilling; check the dates and pressure gauge if fitted?				
Are labels on equipment intact and legible?				
Are symbolic signs intact and visibly displayed?				
Have all laboratory technicians been trained on use of the equipment?				
Other comments:				

Competent person or safety representative: ....

Responsible person or management: ....

Appendix II
Internationally accepted laboratory procedures and test methods

Association of Official Analytical Chemists (**AOAC**): *Official Methods of Analysis*

This is an established international reference for test methods on foods, fertilisers, cosmetics, etc.

It was commonly known as the Association of Official Agricultural Chemists established in 1894. It is now known as the AOAC International.

American Oil Chemists' Society (**AOCS**)

Previously known as the Society of Cotton Chemists Products Analysts, founded in 1909.

This is a comprehensive reference for vegetable fats and oils' test methods and procedures.

American Society for Testing and Materials (**ASTM**)

This is an excellent established set of volumes of international standard test methods covering just about everything.

Pharmacopoeia

British Pharmacopoeia or *United States Pharmacopeia* or *European Pharmacopoeia*, monographs and formulary.

Reference works for pharmaceuticals, cosmetics and nutraceutical methods and specifications.

Standard Methods for the Examination of Water and Wastewater (**APHA**, AWWA and WEF)

This is a well-known reference manual for drinking (potable) water and effluent testing.

International Organisation for Standardisation (**ISO**)

Widely used standardised test methods, standards and specifications are used in trading merchandise and commodities, mainly by European and Western countries.

The *Codex Alimentarius* (Latin for *Book of Food*) is a collection of internationally recognised food standards, codes of practice as well as test methodologies.

The Institute of Petroleum (**IP**)

A source of wealth of information on mineral oils, lube oils, fuels and waxes.

In addition, many other Test Method Standards and Specifications from individual countries are used for import and export quality checks and forex documentation.

https://doi.org/10.1515/9783110721201-028

Appendix III
Strengths of common concentrated acids

Acid	Chemical symbol	% by weight	Density or SG	Approx. strength in mol/L or M	Approx. strength in N	mL acid required to make 1 L ca 1.0 M strength (1.0 mol/L)
Acetic acid, glacial	CH_3COOH	100	1.05	18	18	60
Acetic acid	CH_3COOH	96	1.06	17	17	62.5
Formic acid	HCOOH	98–100	1.22	26	26	38.5
Hydrochloric acid	HCl	36	1.18	12	12	87
Hydrochloric acid	HCl	32–34	1.16	10	10	100
Hydrochloric acid fuming	HCl	37	1.19	12.5	12.5	86
Hydrofluoric acid	HF	48	1.14	28	28	36.2
Hydrofluoric acid	HF	40	1.13	23	23	44.2
Nitric acid, fuming	HNO_3	99–100	1.52	21	21	44.2
Nitric acid	HNO_3	65	1.4	14	14	65
Perchloric acid	$HClO_4$	70	1.67	12	12	84.7
Perchloric acid	$HClO_4$	60	1.53	9	9	108.7
Phosphoric acid, ortho	H_3PO_4	89	1.75	16	48	62.5
Phosphoric acid, ortho	H_3PO_4	85	1.7	15	45	68
Sulphuric acid, fuming (oleum)	H_2SO_4	110 (65%SO_3)	1.99	–	–	–
Sulphuric acid	H_2SO_4	95–98	1.84	18	36	54.6

https://doi.org/10.1515/9783110721201-029

Note the difference between a **strong** acid and a **concentrated** acid.

A strong acid is one that can be easily ionised, e.g. H^+Cl^-; $H_2^{++}SO_4^-$; $H^+NO_3^-$.

A concentrated acid is one that contains maximum amount of substance, e.g. 99% H_2SO_4; 36% HCl.

A **weak** acid is one that is not easily ionised (but is covalent bonded), e.g. CH_3COOH and HCOOH.

Glacial acetic acid is acetic acid that contains no free water, i.e. consider it as a type of "concentrated" acid.

Organic acids are considered to be weak acids because of its covalent bonding.

Inorganic acids are considered to be strong acids because of their ionic bonding.

Appendix IV
Units of measurement and concentration

Weights

1 kilogram (kg)	= 1,000 grams (g)
1 gram (g)	= 1,000 milligrams (mg)
1 milligram (mg)	= 1,000 micrograms (µg)
1 microgram (µg)	= 1,000 nanograms (ng)
1 nanogram (ng)	= 1,000 picograms (pg)
1 ounce (oz)	= 30 gram (g)
1 pound (lb)	= 16 ounces (oz)
1 gram (g)	= 15 grains (apothecary) (approx.)
1 grain (apothecary)	= 60 milligrams (mg) (approx.)

Volume or liquid

1 g/mL	$= 1\,kg/L = 1 \times 10^3\,kg/m^3 = 1 \times 10^2\,kg/hL$
1 g/L	$= 1\,kg/m^3$
1 L	$= 1\,dm^{-3}$
1 L	= 1,000 mL
1 L	= 1 quart (approx.)
250 mL	= 8 fluid ounces
15 mL	= 4 fluid drams (apothecary) (approx.)
1 mL	= 15 minims (apothecary) (approx.)
1 mL	= 1 cubic centimetre (cc) (approx.)

Percentages and concentrations

1% = 1 part per 100 parts = 10 g/kg = 10,000 ppm = 10,000 µg/g = 10,000 mg/kg
0.1% = 1 part per 1,000 parts = 1 g/kg = 1,000 ppm = 1,000 µg/g = 1,000 mg/kg
0.01% = 1 part per 10,000 parts = 0.1 g/kg = 100 ppm = 100 µg/g = 100 mg/kg
0.001% = 1 part per 100,000 parts = 0.01 g/kg = 10 ppm = 10 µg/g = 10 mg/kg
0.0001% = 1 part per 1,000,000 parts = 0.001 g/kg = 1 ppm = 1 µg/g = 1 mg/kg

Parts per million

part per million = ppm = 1 part solute in 1,000,000 parts solution
1 ppm by volume = 1 mg/L = 1 µg/mL = 1 ng/µL = 1% m/v $\times 10^{-4}$
1 ppm by weight = 1 mg/kg = 1 µg/g = 1 ng/mg = 1% m/m $\times 10^{-4}$
ppm by weight = ppm by volume × density NOTE: density = mass/volume

https://doi.org/10.1515/9783110721201-030

Parts per billion

part per billion = ppb = 1 part solute in 1,000,000,000 parts solution

1 ppb by volume = 1 μg/L = 1 ng/mL = 1 pg/μL = 1% m/v $\times$ 10^{-7}

1 ppb by weight = 1 μg/kg = 1 ng/g = 1 pg/μg = 1% m/m $\times$ 10^{-7}

Concentration equivalents

1 litre = 1 cubic decimetre ($1dm^3$) 1 mL = 1 cubic centimetre (1 cm^3) = 1 cc
1 part per million = 1 ppm = 1 part solute in 1,000,000 parts solution
1 microgram/litre (μg/L) = 1 part per billion = 1 ppb (by volume) = 0.001 ppm (by volume)
1 ppm (by volume) = 1 mg/L = 1 μg/mL = 1 ng/μL = 1% m/v $\times$ 0.0001
1 ppm (by volume) = 1 cm^3/m^3 = 1 μL/L
1 ppm (by weight) = 1 mg/kg = 1 μg/g = 1 ng/mg = 1% m/m $\times$ 0.0001
1% = 1 part solute per 100 parts solution
1% m/v = 1 mass part solute per 100 volume parts solution
1% m/m = 1 mass part solute per 100 mass parts solution
10% = 10 parts solute per 100 parts solution
1 gram/tonne (g/T) = 1 part per million (ppm) = 1 milligram/litre (mg/L)

Avoirdupois weights

1 ounce (oz) = 16 drams (dr) = 28.3495 gram
1 pound (lb) = 16 oz = 7000 grains = 454 grams
1 dram (avoir.) = 1.7718 gram
1 stone = 14 lb = 6.35 kilogram
1 quarter = 28 lb = 12.701 kilogram (kg)
1 hundredweight (cwt) long = 112 lb = 50.80 kg
1 hundredweight (cwt)short = 110 lb = 45.36 kg
1 long ton = 2240 pounds (lbs) = 1.016 metric ton
1 short ton = 2,000 lbs = 0.90718 metric tons
1 cental (ctl) = 100 pound (lb)
1 troy ounce (oz t) = 1 apothecaries' ounce

(continued)

1 apothecaries' ounce = 480 grains = 30 grams
1 apothecaries' ounce = 20 pennyweights (dwt)
1 pennyweight (dwt) troy = 24 grains = 1.55517 gram (g)
1 grain = 0.0648 gram (g)
1 dram (apoth.) = 60 grains = 4 g
1 scruple (apoth.) = 1.296 gram
1 drachm (apoth.) = 3.888 gram

Standard concentrations of reagent solutions
- **Normality** = N = number gram equivalent weight of solute per 1 litre solution.
  Where:

$$\text{equivalent weight} = \frac{\text{gram molecular weight of solute}}{\text{valency of reaction}}$$

(Note: the value of "N" is determined by the type of reaction of the titration)
- **Molarity** = M = number moles of solute per I litre solution.

Where

$$\text{mole} = \text{gram molecular weight} = \text{mol}$$

therefore

$$M = \text{mole}/\text{litre} = \text{mol.l}^{-1} = \text{mol.dm}^{-3}$$

(Note: the value of "M" is independent of the type of reaction of the titration)
- **Molality** = m = number moles of solute per 1 kilogram **solvent.**
  (Note: the value of "m" is number moles of solute per kilogram solvent not kilogram solution)
- **Formality** = f = number gram formula weight solute per 1 litre **solution.**
  (NOTE: "f" IS COMMONLY USED IN USA, BUT IN PRACTISE IT IS SAME AS MOLARITY)

Appendix V
International System (SI) of Units

Property	Basic SI unit	Derived units
Length	Meter (m)	Kilometre (km), centimetre (cm)
Mass	Kilogram (kg)	Gram (g)
Volume	Cubic metre (m^3)	Litre (L), millilitre (mL), cubic centimetre (cm^3, cc)
Thermodynamic temperature	Kelvin (K)	Celsius (C)
Energy	Joule (J)	Calorie (c), kilocalorie (kcal, C)
Time	Second (s)	Millisecond (ms)
Electric current	Ampere (A)	Milliamps (mA)
Amount of substance	Mole (mol)	A base unit of SI
Luminous intensity	Candela (cd)	A base unit of SI

Property	Derived unit
Frequency	Hertz (Hz)
Force	Newton (N)
Pressure	Pascal (Pa)
Power	Watt (W)
Electric charge	Coulomb (C)
emf, potential difference	Volt (V)
Capacitance	Farad (F)
Electric resistance	Ohm (Ω)
Electric conductance	Siemens (S)
Electric inductance	Henry (H)
Magnetic flux	Weber (Wb)
Magnetic flux density	Tesla (T)
Luminous flux	Lumen (lm)
Luminance	Lux (lx)
Activity (radionuclide)	Becquerel (Bq)

https://doi.org/10.1515/9783110721201-031

(continued)

Property	Derived unit
Adsorbed dose	Gray (Gy)
Dose equivalent	Sievert (Sv)
Catalytic activity	Katal (kat)
Plane angle	Radian (rad)
Solid angle	Steradian (sr)

Appendix VI
Metric (SI) prefixes and decimal unit multiples

Multiple	Prefix	Symbol
$1,000,000,000,000,000,000,000,000 = 10^{24}$	Yotta	Y
$1,000,000,000,000,000,000,000 = 10^{21}$	Zetta	Z
$1,000,000,000,000,000,000 = 10^{18}$	Exa	E
$1,000,000,000,000,000 = 10^{15}$	Peta	P
$1,000,000,000,000 = 10^{12}$	Tera	T
$1,000,000,000 = 10^{9}$	Giga	G
$1,000,000 = 10^{6}$	Mega	M
$1,000 = 10^{3}$	Kilo	k
$100 = 10^{2}$	Hecto	h
$10 = 10^{1}$	Deka	da
$1 = 10^{0}$	Unit	unit 1
$0.1 = 10^{-1}$	Deci	d
$0.01 = 10^{-2}$	Centi	c
$0.001 = 10^{-3}$	Milli	m
$0.000001 = 10^{-6}$	Micro	μ
$0.000000001 = 10^{-9}$	Nano	n
$0.000000000001 = 10^{-12}$	Pico	p
$0.000000000000001 = 10^{-15}$	Femto	f
$0.000000000000000001 = 10^{-18}$	Atto	a
$0.000000000000000000001 = 10^{-21}$	Zepto	z
$0.000000000000000000000001 = 10^{-24}$	Yocto	y

Note:

$10^{0} = 1$
$10^{1} = 10$
$10^{2} = 100$
$10^{3} = 1,000$
$10^{4} = 10,000$
$10^{5} = 100,000$

https://doi.org/10.1515/9783110721201-032

Appendix VII
Temperature scales and conversions

°C = °Celsius = °Centigrade = (°F − 32) × 5/9
Note that the zero on the Celsius is the ice point 273.15 °K
°F = °Fahrenheit = (9/5 × °C) + 32
°F = (K × 9/5) − 459.67
K = Kelvin = (5/9) × (°F + 459.7)
Absolute zero (K) = −273.15 °C
Absolute scale = Kelvin scale = °C + 273.15
°Réaumur = 5/4 × °C
°Réaumur = (°F − 32) × 4/9
°Rankine = °R = (°C + 273.15) × 9/5
°Rankine (i.e. absolute Fahrenheit) = °F + 459.7

https://doi.org/10.1515/9783110721201-033

Appendix VIII
Vacuum units

(approximate)

---

1 mbar = 1×10^{-3} = 0.75 torr = 1×10^{-3} atm

---

1 bar = 1×10^5 Pa = 1 atm = 14.5 lbf./in^2

---

1 torr = 1.333 mbar = 1 mmHg = 13.6 mmH$_2$O

---

1 Pa = 1 N/m^2 = 0.01 mbar = 7.5×10^{-3} torr

---

1 atm = 30 in Hg = 14.7 lbf/in^2 = 1×10^4 mmH$_2$O

---

1 lbf/in^2 = 2 in Hg = 52 torr = 69 mbar

---

1 kgf/cm^2 = 1×10^4 mmH$_2$O = 14.2 lbf/in^2

---

1 in Hg = 25.4 mmHg = 25.4 torr = 34 mbar

---

1 mmHg = 13.6 mmH$_2$O = 133 Pa = 1 torr

---

1 in H$_2$O = 2.5 mbar = 2.5×10^{-3} kgf/cm^2

---

1 mmH$_2$O = 9.8×10^{-5} bar = 3×10^{-3} in Hg

---

1 mmHg = 0.0394 in Hg = 1.33 kPa

---

1 mmHg = 0.01934 psi

10^{-3} to 10^{-7} mbar = high vacuum

---

1 mbar to 1 bar = medium vacuum

---

https://doi.org/10.1515/9783110721201-034

Appendix IX
Pressure units

(approximate)

1 psi = 1 lbf/in^2 = 144 lbf/ft^2
1 psi = 2.310 ft headwater at 20 °C
1 psi = 0.704 m water
1 psi = 27.7 in water
1 psi = 0.0700 kg/cm^2
1 psi = 2.043 in Hg at 20 °C = 51.88 mmHg
1 psi = 5.188 cmHg = 51.71 mmHg
1 psi = 0.0690 bar = 6.895 kN/m^2
1 psi = 68.947 millibar (mbar) = 6.895 kPa
1 bar = 105 N/m^2
1 N/m^2 = 1 Pascal = 10 dyne/cm^2
1 bar = 14.50 lbf/in^2 = 14.50 psi
1 bar = 1.02 kgf/cm^2 = 0.9869 atm.
1 kgf/cm^2 = 1 kP/cm^2 = 1 technical atm
1 atm = 762.5 mmHg
1 lbf/ft^2 = 0.4788 mbar = 0.3591 mmHg

https://doi.org/10.1515/9783110721201-035

Appendix X
Autoclave steam pressure and temperature

Steam pressure (lbs/sq.in)	Temperature (°C)	Temperature (°F)
0	100	212
2	104	219
5	108	227
8	113	235
10	115	239
12	118	244
15	121	250
18	124	255
20	126	259
22	128	262
25	130	267
28	133	271
30	135	274

https://doi.org/10.1515/9783110721201-036

Appendix XI
Drying agents – desiccants

Desiccant solid	Formula	Water content of air in equilibrium, in mg/L at 25 °C
Copper sulphate	$CuSO_4$	1.4 (least absorbent)
Zinc chloride, fused	$ZnCl_2$	0.8
Sodium hydroxide	$NaOH$	0.4
Calcium chloride	$CaCl_2$	0.2
Magnesium oxide	MgO	0.01
Calcium sulphate	$CaSO_4$	0.01
Potassium hydroxide	KOH	0.01
Calcium sulphate	$CaSO_4$	0.005
Aluminium oxide	Al_2O_3	0.004
Silica gel (blue-dried)	$(SiO_2)_x$	0.003
Sulphuric acid, concentrated	H_2SO_4	0.003
Magnesium perchlorate	$Mg(ClO_4)_2$	0.002
Barium oxide	BaO	0.001
Molecular sieves	–	0.001
Phosphorus pentoxide, flakes	P_2O_5	0.0001 (most absorbent)

https://doi.org/10.1515/9783110721201-037

Appendix XII
Constant humidity solutions

Solute saturated solution	Formula	(Approx.) relative humidity above solution at 20 °C
Lead nitrate	$Pb(NO_3)_2$	98
Sodium carbonate	$Na_2CO_3 \cdot 10H_2O$	92
Potassium chloride	KCl	86
Ammonium sulphate	$(NH_4)_2SO_4$	80
Sodium chloride	$NaCl$	76
Sodium nitrite	$NaNO_2$	65
Ammonium nitrate	NH_4NO_3	63
Calcium nitrate	$Ca(NO_3)_2 \cdot 4H_2O$	55
Potassium thiocyanate	$KSCN$	47
Zinc nitrate	$Zn(NO_3)_2 \cdot 6H_2O$	42
Calcium chloride	$CaCl_2 \cdot 6H_2O$	35
Potassium acetate	CH_3COOK	20
Lithium chloride	$LiCl \cdot H_2O$	15

https://doi.org/10.1515/9783110721201-038

Appendix XIII
Freezing mixtures

In the order of decreasing temperature

Mixture	Approximate temperature (°C)
Ice	0
10 parts of water + 3 parts of ammonium chloride	−3
10 parts of water + 7.5 parts of sodium nitrate	−5
10 parts of water + 3.5 parts of sodium chloride	−10
1 part of ammonium nitrate + 1 part of water	−15
3 parts of crushed ice + 1 part of sodium chloride (salt)	−20
2 parts of crushed ice + 2 parts of calcium chloride hexahydrate	−30
Dry ice + alcohol	−70
Dry ice + chloroform	−80
Dry ice (solid carbon dioxide) + methanol or acetone	−85
Dry ice (solid carbon dioxide) + ether	−100
Liquid nitrogen	−200

https://doi.org/10.1515/9783110721201-039

Appendix XIV
Rate of flow units

1 L/min (lpm) × 0.03532 = 1 cubic foot/minute (cu ft/min)

1 L/min (lpm) = 0.22 gal/min

1 L/min (lpm) × 0.2642 = 1 gal (US)/min

$1 \text{ m}^3/\text{s} = 3.6 \times 10^6 \text{ L/h}$

$1 \text{ m}^3/\text{day} = 0.6944 \text{ L/min (lpm)}$

$1 \text{ m}^3/\text{h} = 0.2778 \text{ L/s}$

$1 \text{ gal/min} = 0.2727 \text{ m}^3/\text{h}$ (cubic metres per hour)

I lpm = 60 L/h (litres per hour)

https://doi.org/10.1515/9783110721201-040

Appendix XV
Detection limits in atomic spectroscopy

Limits (approximate) are given in micrograms per litre:

Symbol	Element	AAS flame (air–acetylene)	AES flame emission (air–acetylene)	AAS graphite furnace	AAS hydride	ICP-OES (axial)
Ag	Silver	2	20	0.01		1
Al	Aluminium	5	5	0.1		1
As	Arsenic	2		0.1	0.03	1
Au	Gold	9		0.2		1
B	Boron	1,000		20		1
Ba	Barium	2	2	0.4		0.03
Be	Beryllium	2		0.01		0.1
Bi	Bismuth	30		0.1	0.03	1
Ca	Calcium	2	0.1	0.01		0.1
Cd	Cadmium	1		0.002		0.1
Ce	Cerium			–		5
Cs	Caesium	15		0.1		–
Co	Cobalt	9		0.2		0.2
Cr	Chromium	3	5	0.01		0.2
Cu	Copper	2	10	0.01		0.4
Fe	Iron	5	50	0.1		0.1
Ga	Gallium	75		–		2
Ge	Germanium	300		–		1
Hg	Mercury	300		1	0.01	1
In	Indium	30		–		1
Ir	Iridium	900		3		1
K	Potassium	3	3	0.01		1
La	Lanthanum	3,000		–		0.4
Li	Lithium	1		0.1		0.3
Mg	Magnesium	0.2	5	0.004		0.04

https://doi.org/10.1515/9783110721201-041

(continued)

Symbol	Element	AAS flame (air–acetylene)	AES flame emission (air–acetylene)	AAS graphite furnace	AAS hydride	ICP-OES (axial)
Mn	Manganese	2	15	0.02		0.1
Mo	Molybdenum	45	100	1		0.5
Na	Sodium	0.3	0.1	0.02		0.5
Ni	Nickel	6	600	0.5		0.5
Pb	Lead	15	200	0.2		1
Pd	Palladium	30		0.1		2
Pt	Platinum	60		2		1
Rb	Rubidium	3		0.03		5
Rh	Rhodium	6		–		5
Ru	Ruthenium	100		1		1
S	Sulphur	–		–		25
Sb	Antimony	45		0.1	0.2	2
Sc	Scandium	30		–		0.1
Se	Selenium	100		0.1	0.03	2
Si	Silicon	90		1		10
Sn	Tin	150	300	5		2
Sr	Strontium	3		0.03		0.1
Te	Tellurium	30		0.1	0.03	2
Ti	Titanium	75		0.4		0.4
Tl	Thallium	15		0.1		2
Tm	Thulium	15		–		1
V	Vanadium	60	200	1		0.5
Yb	Ytterbium	8		–		0.1
Zn	Zinc	2	200	0.02		0.2

*AAS limits are for the air/acetylene flame at 100 mm path length and also depend upon the model and manufacturer of the instrument.

Appendix XVI
Oxidation states

Element	Group	Atomic number	Oxidation state
Li	1 (IA)	3	**+1**
Be	2 (IIA)	4	**+2**
B	13 (III)	5	**+3**, −1, +1, +2
C	14 (IV)	6	**+4, −4**, +2
N	15 (V)	7	**−3**, +2, +3, +4, +5
O	16 (VI)	8	**−2**, −1, +1, +2
F	17 (VII)	9	**−1**
Na	1 (IA)	11	**+1**, −1
Mg	2 (IIA)	12	**+2**, +1
Al	13 (III)	13	**+3**, +1
Si	14 (IV)	14	**+4, −4**, −1, −2, −3, +1, +2, +3
P	15 (V)	15	**−3, +3, +5**, +4
S	16 (VI)	16	**−2**, +2, +4, +6
Cl	17 (VII)	17	**−1**, +1, +3, +5, +7
K	1 (IA)	19	**+1**
Ca	2 (IIA)	20	**+2**
Cr	6 (VI)	24	**+3, +6**
Mn	7 (VII)	25	**+2, +4, +7**
Fe	8 (VIII)	26	**+2, +3**, +4, +6
Co	9 (VIII)	27	**+2, +3**
Ni	10 (VIII)	28	**+2**
Cu	11 (IB)	29	**+2, +1**, +3, +4
Zn	12 (IIB)	30	**+2**
Mo	6 (VI)	42	**+2, +4, +6**
I	17 (VII)	53	**−1**, +1, +5, +7
Pb	14 (IV)	82	**+2, +4**
Bi	15 (V)	83	**+3**, +5

https://doi.org/10.1515/9783110721201-042

Appendix XVII
Commodity properties

Commodity	Density or SG (kg/m³)	Auto-ignition (°C)	Flash point, (°C)
Asbestos	320–400	Not combustible	Not combustible
Charcoal	400	350	Undetermined
Cement	1,290–1,600	Not combustible	Not combustible
Coal, bituminous	760–840	400	130
Coal, anthracite	840–970	450	Undetermined
Coke (petcoke)	570–650	600–800	100
Cotton	1,550	260	210
Diesel fuel, petr.	810–960	210–400	69
Fuel oil	800–900	200	100/300
Gas oil	800–900	300	30/50
Kerosene	800–900	230	35/80
Lube oil	800–900	400	180/280
Natural gas	700–800	480	−180
Petrol/gasoline	700–800	280	−43
Paper	80–120 grammage	300	230
Peat, hay	400	150–220	Undetermined
Quartz, sand	1,450–2,070	Not combustible	Not combustible
Rubber	1,500	220–300	170
Salt	1,130–1,290	Not combustible	Not combustible
Steel	7,850–7,900	Not combustible	Not combustible
Sulphur	800–1,130	230	200
Wood, shavings	210–520	150	100
Wood, solid	400–1,200	350	300
Wool, natural	1,310	600	Undetermined

https://doi.org/10.1515/9783110721201-043

Appendix XVIII
Properties of metals and alloys

Metal/alloy	Density (SG)	Melting point (°C)
Aluminium (Al)	2.7	650
Antimony (Sb)	6.7	630
Bismuth (Bi)	9.8	270
Brass	8.4	950
Bronze	8.0	1,000
Cast iron	7.5	1,200–1,350
Cadmium (Cd)	8.6	320
Copper (Cu)	8.9	1,080
Cobalt (Co)	8.7	1,480
Chromium (Cr)	7.2	1,880
Gold (Au)	19.3	1,060
Iron (Fe)	7.8	1,530
Lead (Pb)	11.3	330
Manganese (Mn)	7.4	1,250
Mercury (Hg)	13.5	−39
Molybdenum (Mo)	10.2	2,500
Nickel (Ni)	8.8	1,450
Platinum (Pt)	21.4	1,770
Steel	7.8	1,100–1,600
Silver (Ag)	10.5	960
Sulphur (S)	2.1	110
Tin (Sn)	7.3	230
Tungsten (W)	18.8	3,400
Zinc (Zn)	7.2	420

https://doi.org/10.1515/9783110721201-044

Appendix XIX
Care of platinum apparatus

These should only be cleaned with a general household detergent that does *not* contain ammonia.

Platinum (aka white gold) is a very expensive malleable noble metal and must be treated with care. Platinum crucibles and their lids are delicate and can become dented, cracked, pin-holed or damaged in many ways. They should not be heated above 1,200 °C, unless the crucible is a special alloy (usually with gold, rhodium, tantalum or iridium) that can withstand higher **temperatures**.

Any heating above 550 °C (or ignition in muffle furnace) of the apparatus should **not be done under oxidising** atmosphere.

This apparatus must *not* come in contact with iron (e.g. wire gauzes and wire triangles), phosphoric acid, aqua regia, nitric acid, sodium peroxide, ammonia, chlorine and sulphur gases; or presence of any nitrates or nitrites.

Otherwise, platinum is relatively chemically unreactive.

Platinum crucibles and dishes should be **stored** in their original **"formers"** to prevent distortion of the platinum apparatus because platinum is very malleable and ductile.

Platinum ware can be cleaned in boiling hydrochloric acid for about 2 min. The crucibles may also be cleaned by doing a blank fusion with sodium hydrogen sulphate, followed by dilute acid wash.

Cleaning labware
Laboratory procedures require clean labware because the most carefully executed analysis of a sample or an item will yield erroneous results, if contaminated or if dirty glassware was used.

Glassware must be physically (visually) and chemically clean without any non-visible detergent residues present. Some common detergents contain borax, phosphates and alkali metals, which will interfere with many testing procedures leading to false results.

A good indication of cleanliness is **uniform wetting** of the surface of the glass by distilled water. This is very critical in glassware used for dispensing accurate volumes of liquid such as burettes and pipettes. Grease (such as lab grease, desiccator or vacuum grease and Apiezon) and other oils contaminating materials will prevent the glass from becoming uniformly wetted, which in turn will cause liquid "globules" adhering to the glass or plastic surface, thus causing an inaccurate volume of the liquid to be dispensed.

There are many detergents and cleaning agents available; however, **chromic acid** (although **dangerous** to handle) is still the most efficient glassware cleaning chemical; alternatively, one may try a relatively mild "alcoholic caustic".

https://doi.org/10.1515/9783110721201-045

The **type** of cleaning chemical to use is very much dependent upon the type of soiling of the glassware or apparatus; there is no such thing as the "ultimate clean-all universal" agent.

A gentle cleaning paste is toothpaste. It has been used with success for the external cleaning of solid-state ion-selective electrode, pH and reference electrodes, including many stainless parts.

Formulae for cleaning glassware:

Formula for chromic acid (HAZARDOUS CORROSIVE CHEMICAL):

Ideal for inorganic dirt, mineral residues and calcium hardness deposits (scale).

Approximately 25 g of sodium dichromate is dissolved in 25 mL water and add slowly concentrated sulphuric acid with stirring to the final volume of about 1 L; note the evolution of heat.

Formula for alcoholic caustic solution (HAZARDOUS CHEMICAL):

This is ideal for organic dirt, fats, oils, foodstuffs and similar greases.

Approximately 56 g of potassium hydroxide is dissolved in minimum water, and then add alcohol (methylated spirits) to make up to about 1 L (i.e. ±1 N alcoholic KOH). The cleaning reaction is saponification of fats into soaps.

Caution: do not soak glass items for longer than 15 min in this solution.

Formula for *extra* effective general detergent (HAZARDOUS ACID BURNS):

Add about 10 mL of hydrofluoric acid to about 1 L general household dishwasher liquid. Mix well.

Caution: do not soak glass items for longer than 10 min in this solution because the HF acid attacks glass (silicates) and any markings on the glassware.

Formula for cleaning sintered glass filter crucibles (HAZARDOUS ACID BURNS):

Boil gently the crucibles in aqua regia (3 parts hydrochloric acid + 1 part nitric acid) for about 5 min in a fume cupboard. Then flush crucible well with water whilst under vacuum filtration.

Formulae for disinfecting and sanitising bench surfaces, work areas and hands:

- 70% v/v ethyl alcohol (aqueous blend); or
- 75% v/v alcohol + 2% v/v ether + 1% v/v acetone + balance water. Do not apply to plastics. Either ethyl alcohol or isopropyl alcohol may be used; or
- 80% v/v ethyl alcohol + 1.4% m/v glycerine + 0.05% v/v hydrogen peroxide (of 3% H_2O_2 = 10 volumes of O_2); or
- 75% v/v isopropyl alcohol + 1.4% m/v glycerine + 0.05% v/v hydrogen peroxide (of 3% H_2O_2 = 10 volumes of O_2).

Appendix XX
Periodic table

Fig. A.XX: Periodic table.

https://doi.org/10.1515/9783110721201-046

Glossary of terms

This glossary contains terms used in the text together with other words that might be of interest to the student of analytical chemistry and science.

Absolute viscosity (in centipoises) is equal to **kinematic viscosity** (in centistokes) multiplied by **density**.

Acids *and bases* are two classes of chemical compounds that display generally **opposite characteristics**. Acids taste sour, turn litmus (a pink dye derived from lichens) red and will react with metals to produce hydrogen gas. A base (alkali) tastes bitter, turns litmus blue and feels slippery to the touch. When aqueous (water) solutions of an acid and a base are combined, a neutralisation reaction occurs. This reaction is characteristically very rapid and generally produces water and a salt.

For example, sulphuric acid (H_2SO_4) and sodium hydroxide (NaOH) yield water (H_2O) and sodium sulphate (Na_2SO_4):

$$H_2SO_4 + 2NaOH = 2H_2O + Na_2SO_4$$

(a stoichiometric equation)

1 molecule H_2SO_4 + 2 molecules NaOH = 2 molecules H_2O + 1 molecule Na_2SO_4

Acid is any substance that is capable of **donating** a proton and a base is any substance that can accept a proton. This is the **Bronsted–Lowry** theory, or proton theory. A substance that **accepts** electrons is known as the **Lewis** theory.

Accuracy of an analytical method is the extent to which test results generated by the method and the true or accepted known value **agree**. It can also be described as the closeness of agreement between the values measured and that of a reference value; this means that an analyst cannot have accurate analytical results without having precision.

Aliquot is a portion (usually a **specific volume** of a liquid) taken for analysis, which is known to be a fraction of the whole sample.

Alkali is a water-soluble **base** yielding a **caustic** solution of pH greater than 7.

Alkaloid is an **organic nitrogen** base that usually occurs naturally in plants and has a powerful action on humans and animals, when digested; e.g. belladonna. These compounds are also manufactured synthetically

Allotropes of an element are the **different physical forms** (crystalline or molecular) of that element; e.g. carbon has two allotropes, namely diamond and graphite.

Alloy is a synergistic **homogeneous** mixture (actually a solution) of two or more metals imparting **specific** metallic properties.

Amperometry is a category of electroanalytical methods used in chemistry and biochemistry for the detection of ions in a solution based on **electric current** or changes in electric current.

Amphiprotic solvent is a substance that exhibits both acidic and basic properties; e.g. water:$H_2O = H^+ + OH^-$

Amorphous means **without crystalline** or regular atomic structure or shape; e.g. glass, rubber, plastics and gels.

Amphoteric substances are those that can act as **both** an acid or a base.

https://doi.org/10.1515/9783110721201-047

Analyte (aka measurand) is the element or compound that is of **interest** in the measurement or analysis.

Anhydrate is a hydrate molecule that has lost its water molecule, compared to a substance that does not contain any water and is referred to as **anhydrous**. A **hydrate** is a solid compound containing or linked to water molecules. An **anhydride** is any compound formally derived from another by the loss of a water molecule; basically, this is a molecule with no water.

Anode is the electrode at which oxidation occurs.

Apothecary units are used in the pharmacy field and are closely related with the English **troy** system of weights.

Assay is a laboratory (and metallurgy) process used to **analyse** and determine the proportions of precious metals (e.g. gold and silver) in ores and minerals and the purity of pharmaceutical products whose assay methods are often referenced in various pharmacopeia.

Atom is the **smallest part** of a specific element. Atoms of the same element (e.g. sodium) are identical in physical and chemical properties and have the same atomic number (the number of electrons or protons). An atom consists essentially of a nucleus (containing protons and neutrons) surrounded by electrons.

Atomic fluorescence spectrometry is where characteristic fluorescent radiation is emitted from atomic vapour of an analyte following irradiation with UV–Vis primary radiation from a broad spectrum.

Atomic mass of an atom is the **number** of **protons** and number of **neutrons**. The atomic mass used in calculations and given in the periodic table is a weighted average (as found in nature) of the atomic masses of the isotopes of the element.

Avoirdupois weight unit is a system of weights **based on the pound** (lb), which contains 16 ounces or 7,000 grains. About 100 pounds (US) or 112 pounds (British) is equal to 1 hundredweight and 20 hundredweights equals 1 ton.

Aqua regia is a **mixture** of nitric and hydrochloric acids (1:4). It dissolves the noble metals (gold and platinum), hence, its name (aqua means water and regia means royalty).

Azeotrope is a mixture of liquids that **boils** at a **constant temperature**; e.g. a blend of water and xylene.

Background noise is the random or consistent signal noise or fluctuations, causing errors in the measurement of electrical or mechanical outputs, due to environment or instrumental factors.

Becquerel is a unit, Bq, of **radioactivity** corresponding to one nuclear transition per second.

Beer's law is the functional **relationship** between the quantity measured by absorption (A) of a light beam to that of the quantity sought of an analyte (concentration, c), namely $A = \alpha bc$, where α is the proportionality constant known as **absorptivity** and b is the thickness of the absorbing substance (usually the cell length in absorption spectrophotometry).

Bias is the tendency of a method towards delivering a result that is **skewed** from the true value. It is the difference between the experimental mean and the true value.

Billion is a million million in the SI system of units, i.e. it is 1,000,000,000,000. But in the USA, a billion is 1,000 million, i.e. 1,000,000,000.

Birefringence is a type of **double refraction** of light, where a polarised light beam propagates through a material having a particular refractive index, which exhibits characteristic properties of that material. This phenomenon is used widely in identification of minerals and gemstones.

Blank is a prepared solution of all the reagents and processes used in a particular analysis, minus the presence of the sample itself. It is tested and measured at the same time under the same conditions as the sample. The purpose is to correct for any background or chemical interferences during the course of the analysis.

Buffer solution is a mixture of substances that tend to hinder any large **changes** in acidic or basic properties of the solution.

Calibration is undertaken when a response from an instrument's (or volumetric glassware) output is measured, or checked, against a response from a substance (chemical standard or as sized glassware) of known concentration, under the same instrumental conditions.

Cathode is the electrode at which reduction occurs.

Caustic is a substance that is said to cause corrosion (by alkalies) of metals or destruction of organic matter; chemically sodium hydroxide (caustic soda) is considered to be caustic; also anything that has alkaline properties (**pH between 7 and 14**) may be considered to be caustic.

Centrifugal force is a parameter stated when separating substances by using a centrifuge. It can be referred to as **relative centrifugal force** (rcf), where rcf = $(11.18 \times 10^{-6}) \times R \times N$; here R is the radius of the centrifuge head and N is the rpm.

Chromophore is a part of a molecule or compound that shows a **characteristic colour** when exposed to light of certain wavelength/s.

Chromaticity is the measurement of the quality colour of an item **independent** of its brightness or hue.

Chromatography is a technique used for **separating** and **identifying** various anions, cations and fractions of organic compounds; e.g. IC, TLC, HPLC, GC-MS.

Coefficient of variance (of a mean) is a statistical representation of the **precision** of a test result. The function: (standard deviation/mean) × 100% = CV.

Colligative properties are that of a solution which depends on the **number**, and not nature, of dissolved **particles**.

Compound is made of elements in **fixed proportions** (ratios); e.g. NaCl and Na_2CO_3. There are billions of natural compounds and new synthetic compounds are being made every day.

Concentration is the amount of a substance present in a given mass or volume of another substance as a whole.

Constituent is a finite part of a substance.

Coulometry is where an **electrode** is **generated** from chemical **reagents** by electrolysis (an applied voltage potential between an electrode and its generated counter electrode). Refer to Karl Fischer coulometric titrations.

Deliquescence is where solid substances **absorb water** from the air and dissolve in it to form a liquid.

Denature is when the tertiary structure of a **protein** breaks due to heating, change in pH, agitation or even exposure to air.

Density is by definition, **mass per unit volume**. There are three types of density: density in vacuo, true density and apparent density. The former two are tested by hydrometer or an automatic density meter and the latter one is tested by pyconometer or SG bottle (also known as a Gay–Lussac bottle). The units for density are in mass/volume and at a specific temperature, e.g. kg/L at 20 °C.

Detection limit is the smallest amount or concentration of an analyte (e.g. an element, compound, molecule and functional group) that can be detected by a given analytical procedure and with a stated degree of confidence (e.g. 95% confidence level).

Determination is a reported quantitative measure of an analyte with a stated degree of accuracy, or a stated factor of uncertainty of measurement (UoM).

Determinate and *indeterminate error*, where the **former** are errors that can be **identified** and possibly reduced or eliminated and the **latter** are just **inherent** errors that appear naturally from time to time and cannot be measured or controlled.

Desiccants are substances that are used to keep **items dry**; they absorb moisture by either chemical reaction (absorption), e.g. phosphorus pentoxide changes into phosphoric acid; or by physical adsorption (e.g. silica gel).

Distilled water is pure water made by boiling the water and **condensing** the steam; this should preferably be done in an all-glass water still, alternatively a stainless steel water still. The water contains no dissolved gases but can absorb carbon dioxide from the atmosphere and could also contain organics. However, this water could be high in silicates due to the glass apparatus. Sometimes the water is triple distilled to obtain a conductivity of less than 1 microSiemen per centimetre (<1 µS/cm at 25 °C).

De-ionised water is pure water made by passing water through **ion-exchange resins** to remove soluble inorganic salts; however, it does not remove organic impurities and dissolved gases in the water. This type of water should not be used for microbiological tests. For ultra-pure water (e.g. Type I), the water can be first distilled than passed through a de-ioniser, or via a reverse osmosis membrane.

Electrochemical cell is a pair of electrodes (usually a non-corrosive metal such as platinum) in contact with an electrolytic medium (usually a solution).

Electrode potential is the potential (voltage) measured relative to a standard, usually the **standard hydrogen electrode** (SHE).

Electrolytic cell is an electrochemical cell through which current is forced (electrolysis) by a battery or some other external source of energy.

Electrolysis is where **electricity** is carried or passed, **through a solution** by ions, from one electrode or terminal (anode) to another electrode or terminal (cathode).

Electrolyte is a strong electrolyte when it can almost dissociate completely into ions in a solution, whereas a weak electrolyte is one that remains mostly undissociated in a solution.

Electromotive series or *electrochemical series* is a series (list) of elements arranged in order of their electrode potentials (or **ability to replace metals** from one another when in their salt form). The series is K, Ca, Na, Mg, Al, Zn, Cd, Fe, Ni, Sn, Pb, H_2, Cu, Hg, Ag, Pt and Au (where H_2 is taken to be zero). Thus, a copper salt will replace a zinc metal.

Electrophoresis is the **separation** and migration of charged particles such as ions under the **influence of an applied electric field** in an electrolytic medium (paper, cellulose acetate or polymeric gels). The migration rates depend upon the pH and temperature of the medium, the size, shape and electric charge of the species. It is used extensively in the field of nucleic acids and amino acids (proteins).

Element is **made of atoms**; there are 118 (or more) elements known and these are listed in a chart (known as the periodic table), in order of their specific chemical and physical properties; e.g. H, He,

Li, Be, B, listing down and across as one would normally read a book. Elements in a column have similar properties and elements in a row also have similar specific type of properties.

Enantiomers are chiral molecules that are **mirror images** of one another; e.g. dextro-lactic acid and laevo-lactic acid.

End point is the point in the progress of a reaction (e.g. titration), which may be precisely located (identified and detected) and which can be related to the stoichiometric or equivalence point of that chemical reaction.

Equilibrium refers to reactions in which the forward and reverse rates are matched where the composition of the mixture appears to be **unchanging** in a specified time period.

Equivalent is the amount of a substance which will, in a chemical reaction, be equated to 1 mol of hydrogen ions (6.023×10^{23}).

Error is the difference between a measured value and the **true** or established or most probable value.

Estimation is a semi-quantitative measure of the amount of an analyte present in a sample; usually having a degree of accuracy not better than 10%.

Fire point of a liquid indicates the lowest temperature at which the solid or liquid's vapour would *continue to burn in air*, when exposed initially to a source of flame and the flame is then removed. A special flash point apparatus of international standard dimensions has to be used.

Flash point (closed cup or open cup) indicates the lowest temperature of a liquid at which a flash would occur at a specific temperature when exposed to a source of flame; the flash would automatically burn out and *not* continue to burn. A special flash point apparatus of international standard dimensions has to be used.

Freezing point depression is the process where the **addition of a solute** to a solvent **decreases** the freezing point of the solvent. Example: addition of 10 g salt (solute) to 100 mL water (solvent) causes the solution to freeze at −6 °C.

Eutectic is a **mixture** of two or more substances at a composition yielding the **lowest melting point**, in contrast to dystectic which is the composition yielding the maximum melting point.

Fluorometry is an analytical technique that uses fluorescence to detect and identify small samples of a substance.

Functional groups are the identifiable groups of atoms exhibiting **characteristic properties** or reactions of a compound; e.g. alcohols have the −OH group.

Fusion is the **melting**, at high heat of about 900 °C, of one or more inorganic salts (flux) with the insoluble substance (sample) to form a uniform single substance which then becomes **soluble** in a liquid reagent; e.g. the heating of sand (silica) with soda ash (sodium carbonate) to form water-soluble silicate glass.

Fluorescence is a property of some atoms or molecules when there is a rapid **emission of light at longer wavelengths** than that which is absorbed; e.g. adsorption of blood under UV light.

Flux (derived from the Latin fluxus meaning "to flow") is a substance, usually an inorganic compound, added to a refractory substance under high temperatures, to **lower** the mixture's **melting point** so as to form a fusion or melt; e.g. glass. Examples of fluxes used in the laboratory are sodium hydroxide pellets, sodium carbonate, sodium peroxide, potassium pyrosulphate, lithium metaborate and lithium tetraborate.

Galvanic cell (voltaic cell) is an electrochemical cell that spontaneously produces current (energy 0 when the electrodes are connected externally by a conducting wire).

Gel permeation chromatography is a process where molecular aggregates of the sample are **separated** in accordance with their **molecular size**.

Half-cell reaction is a reaction when oxidation or reduction occurs at one of the electrodes in an electrochemical cell.

Heterogeneous matter is **not uniform** in composition but consists of two or more physically distinct properties. Each physically distinct homogeneous portion of a heterogeneous mixture is called a phase; e.g. yoghurt containing fruit particles and pure orange juice with pulp.

Heavy metals are those elements generally considered to have a **density greater than 4** (compared to water that has a density of 1), such as Pb, Zn, Cu, Cd, Cr, Hg, Fe, Pt and Ag. Basically, metals with higher atomic mass tend to form compounds that are more poisonous than those compounds containing elements (metals) of lower atomic mass.

Homogeneous matter is **uniform throughout** its composition with respect to its chemical and physical properties; e.g. salt, sugar, flour and cement.

Homologous series is a series (list) of related compounds that have the **same functional group** but differ in their formula by a fixed group of atoms; e.g. carboxylic acids such as formic acid (HCOOH), acetic acid (CH_3COOH) and propionic acid (C_2H_5COOH).

Humectants are additives or compounds that tend to **keep** a substance or product **moist**; these are opposite to desiccants.

Hygroscopic are substances that **absorb moisture** from the air (environment), whereas hydroscopic means observing objects below the surface of water. Compare with deliquescence, a process in which a hygroscopic solid substance absorbs moisture from the atmosphere to such an extent that a concentrated liquid is eventually formed; e.g. ammonium acetate.

Indicator is a reagent or device used to indicate when the end point (equivalence point) of a chemical reaction has been attained.

Interference is an effect that alters or obscures the behaviour or measurement of an analyte determination during an analytical procedure.

Internal standard is a compound or element that is added to all calibrations standards and the sample, in a known constant amount. This analytical procedure is usually utilised to determine if any measurement errors are apparent; or as a reference point to measure retention times in chromatography.

Iodometry is an **indirect** titration test procedure in which **oxidising** agents are determined by reacting with an excess of iodide, where the iodine liberated is titrated with a standard reductant such as a solution of sodium thiosulphate, usually in a slightly acidic media.

Iodimetry is a **direct** titration test method in which a standard solution of iodine is used to titrate strong **reductants**, usually in neutral or slightly acidic solution.

Isotopes of an element have **same number** of protons, but different number of neutrons, hence different atomic masses.

Ligand is a species with at least one **Lewis base** (denotes electrons) site which can participate in complex formation. Compare with chelating agents.

Limit of quantitation (LOQ) means the **lowest** concentration level that can be **quantified** with acceptable precision and accuracy, by a measuring instrument, such as a spectrophotometer.

Limit of detection (LOD) is the **lowest** concentration level that can be **detected** but **not quantified** and is generally 50% of the LOQ.

Linearity is the ability of a test method to obtain test data **directly proportional** to the **concentration** of the analyte; also, sometimes referred to as the acceptable working range of a calibration curve or graph.

Luminescence is the emission of light by a substance by any reason other than a rise in its temperature. Generally there are two types, namely **fluorescence** and **phosphorescence**, where both have the ability of a substance to **absorb** light and **emit** light of a longer wavelength and therefore of a lower energy. Phosphorescence is where the light energy is produced by a chemical reaction, which lasts only for a very short period of time in seconds, whereas fluorescence is where the emission of light that has been absorbed by some form of electromagnetic radiation.

Masking is the practice of treating a sample with a reagent to prevent any interference or unwanted side reactions affecting the response of an analyte due to the presence of other constituents of the sample.

Matrix represents the other constituents of a sample composition that is not part of the sought analyte.

Method is synonymous with the term **measurement** procedure. It is the overall description of testing instructions of an analytical procedure.

Meniscus is the **curved surface** of a liquid when the liquid is in a narrow tube such as a burette or pipette. The reading is usually taken on the bottom of the curve. Note that liquid mercury meniscus curves upwards.

Miscible is where two liquids mix **completely** in all proportions to form one phase or solution.

Mnemonics is a common **aid** to use to remember things where words are phrased into an easy sentence of which the first letters of each word (namely an acronym) are intended to remind the person of an event, definition, etc.

Moisture is the **free water** content of a substance (*not* the water of crystallisation of a chemical salt or crystal).

Molecule is the **smallest part** of a compound and is formed by the combination of two or more elements; e.g. a molecule of water is H_2O.

Mossbauer absorption spectroscopy is where a solid sample is exposed to a beam of **gamma radiation**.

Nephelometry is the measurement of the intensity of the scattered light at **right angles** to the direction of the incident light as a function of a turbid liquid.

Nesslerimetry is a form of colorimetry where a **comparison** of two colours of two solutions (liquids) in Nessler tubes or cells are made; e.g. the Lovibond Comparator®. The comparison may also be made visually with the naked eye without the use of any type of apparatus.

Neutron is an electrically **neutral** part of the nucleus of an atom.

Nomogram is a **diagram** representing the **relationships** between three (or more) variable quantities (such as pH, alkalinity and free carbon dioxide in waters), by means of a number of linear (or graphical) scales, so arranged that the value of one variable (commonly the centre scale) can be deduced.

Normality, *molarity, molality and formality* are the various strengths of **standardised** solutions.

Osmosis is where a solvent molecule passes through (diffuses) a **semi-permeable membrane**, due to concentration gradient, but blocks out a solute molecule. Reverse osmosis is where the passage due to pressure exerted by a pump forces the passage or diffusion from lower concentration to a higher concentration level.

Periodic table is a **list** or chart of all the known elements in numerical order according to their **atomic numbers**.

p-Functions is the expression for the concentration of a species in dilute solutions. It is defined as the **negative logarithm,** to the base 10, of the molar concentration of that substance; e.g. $pH = - \log_{10} \{H^+\}$.

Peltier cooling or thermoelectric effect is the reverse of the Seebeck effect; here the electrical current flowing through the junction of two connecting materials (conductors) will emit or absorb heat per unit of time at the junction to balance the difference in the chemical potential of the two materials; that is, heat is removed at one junction and hence cooling occurs. Note that a voltage is applied across the joined materials to create the electric current.

pH is an **arbitrary unit** of acidity or alkalinity on a scale of 0 to 14, where 0 to 7 is increasing in acidity (decreasing in alkalinity) and 7 to 14 is decreasing in acidity (increasing in alkalinity). By definition, pH is the log of the hydrogen ion concentration:

Neutral $pH = 7.0 = 10^{-7}$ g H^+ per litre concentration.

Polarography is a subclass of **voltammetry** where **electrode processes** by means of electrolysis with two electrodes, one polarisable (known as the dropping mercury electrode) and one non-polarisable.

Most common uses are trace quantitation of heavy metals, determination of ionic species of an element (e.g. Fe^{2+} and Fe^{3+}) and some applications in organic analysis.

Polymorphism is the ability of a substance that can be found in a number of **different forms**.

Precision of a determination is measured by the calculated or statistical evaluation of the variability of a test result; usually the methods used are standard deviation or relative standard deviation (aka coefficient of variation (CV) and expressed as RSD%).

Primary standard (aka CRM, that is a certified reference material) is a substance whose purity and stability (i.e. shelf life) are well established and with which other standards may be compared or calibrated against.

Procedure is a description of the practical steps involved in an analytical method, analysis or methodology.

Proton is an electrically **positive part** of the nucleus of an atom.

Qualitative analysis involves determining **what** elements or compounds are present in a substance.

Quantitative analysis involves determining the quantities (**amount**) of each element or compound present in a substance.

Reagent is something (a chemical liquid or powder) that will **react** with something else (a chemical liquid, or powder) to produce a product; usually in an analytical procedure.

Reflectance is defined as the proportion of perpendicularly incident light reflected from a substance to that reflected from another substance of known reflectance.

Relative density is same as **specific gravity**. It has no units as it is a ratio of two densities, usually the density of the substance relative to that of density of water (SG of water = 1.0).

Repeatability is the **closeness** of agreement between the results of successive measurements of the analyte (or measurand) under the same conditions of measurement.

Response factors are used commonly in **chromatography** and sometimes in spectrophotometry; it is the ratio between a detector response signal produced by an analyte and the quantity of analyte which produced that signal.

Reproducibility represents a replicate analysis that is performed using the same methodology on identical (representative) samples, but by **different analysts** or by **different laboratories**.

Robustness is a measure of an analytical procedure's capacity to remain **unaffected** by small deliberate variations in a method's parameters (e.g. different sample weights, dilutions and matrix) only and still provides acceptable precision and accuracy of the final test result.

Sample is the substance or portion of a substance (or statistical population) about which an analysis or evaluation is to be determined.

Selectivity (or *specificity*) is the ability of a test method to determine accurately and **specifically** the **analyte** of interest in the presence of other components in the sample matrix. This term is also sometimes referred to as the analytical specificity.

Sensitivity is how a **change** in an instrument measuring response responds to a change in the analyte **concentration**. The sensitivity is equal to the slope of the standard calibration curve and is a constant value if the curve is linear.

Size exclusion chromatography (aka gel-permeation) is the separation of materials according to their molecular size and shape by passage of a solution through a column or across a surface consisting of a polymeric gel. These gels can be hydrophobic, hydrophilic, rigid silica gels or special types such as Zorbax.

Spectrophotometry is the **measurement** of the transmitted light after passing through a media (sample solution) at a **specific wavelength**. This is used in qualitative and quantitative analyses.

Standard is a pure substance that reacts in a quantitative stoichiometric manner with the analyte or a reagent.

Standard addition is a method or technique used in analytical chemistry whereby the response from an analyte is measured before and after adding a known amount of that analyte to the sample. This technique is often used to eliminate any possible (undetermined) errors or to reduce the detection limit of the analyte under normal conditions of measurement.

Standard deviation is the square root of the average of the squares of deviations about the mean of a set of data. Thus, standard deviation(s) is a statistical measure of **spread** or variability of a set of data.

Standard electrode potential (E°) is the electrode potential measured in solutions where all reactants and products are at unit activity.

Standardisation is a procedure whereby the determination of the concentration of an analyte or a reagent solution, which is compared to that of a standard.

Stoichiometry is the relative **proportions** in which elements form compounds or ratios in which substances react with one another; e.g.

$$MgCl_2 + 2AgNO_3 = 2AgCl + Mg(NO_3)^2$$

Specific gravity (SG) is a **ratio** of **density** of a substance (at temperature T1) to the density of another substance, usually water (at temperature T2). Thus, the SG of a substance at T1/T2 is numerically same as a density of that substance at T1 if the other substance is water at T2=4 °C). SG has no units as it is a ratio, but it is always reported at two temperatures, e.g. 20 °C/4 °C. Note that the density of water at 4 °C is exactly 1.000000. . ..

Sublimation is the transition of a **solid directly** to a **gas** state, without forming a liquid first.

Technique is the principle upon which a group of methods is based, e.g. chromatography.

Trueness is the closeness of agreement between the average of a finite number of replicates of the measured quantity of an analyte and the average of a finite number of replicates of the measured quantity of a reference or control sample. Thus, trueness is the opposite (i.e. inverse) to that of a systematic measurement error or bias.

Thixotropy is the property of becoming **less viscous**, for example on stirring or change on shear rates during mixing of a liquid or paste.

Titration is the overall procedure for the determination of the stoichiometric (equivalence) point of a chemical reaction by various recognised (and well-documented) analytical means.

Titrant is the solution added (e.g. via a burette) or a reagent generated (e.g. coulometric) in a titration.

Titrand is the solution (analyte, in a flask or beaker) to which the titrant is added.

Titer (aka titre) is the concentration of a substance as determined by a titration. That means, the **least volume** (at the end point) needed to complete the reaction in a **titration**.

Titrate is to determine the strength (concentration) of a solution (in a conical flask) by determining the **volume** of a **standard** solution (in a burette) with which it reacts (in a **stoichiometric** relationship).

Triturate is mixing by rubbing, crushing, grinding into fine powder or a **paste**.

Turbidimetry is the measurement of the intensity of the **transmitted** light through a **suspension**, as a function of the concentration of a turbid liquid.

Ultraviolet wavelength range is generally considered to be from about **380 nm** down to **190 nm**. The germicidal (sterilisation) wavelength (UV rays that kill germs) is at 254 and 265 nm. Sometimes the UV range is referred to as UVA, UVB and UVC. The UVC (222 nm) is said to do the sterilisation of bacteria (bactericide) and/or mould and algae (fungicide).

The "Black Box" range for forensic investigations for example, urine or blood detection, is at about 365 nm.

Uncertainty of measurement (UoM or measurement uncertainty) is the parameter associated with the test result that characterises the dispersion of the test values that could reasonably be attributed to the measurand (final test value or reported laboratory analytical result). **Measurement uncertainty** is strictly not a performance characteristic of a particular measurement procedure, but a property of the results obtained using that measurement procedure or process.

Validation of a test **method** or procedure ensures that that process satisfies **all stated** functionality, operations and outcomes with respect to its intended use. This is done by analysing standards that have an accepted analyte content and a matrix similar to that of the sample. Other procedures are the participation in proficiency schemes (aka PT schemes) or round-robins testing (inter-lab comparisons).

Valency is the **combining power** of an atom or radical, equal to the number of hydrogen atoms that the atom could combine with or displace in a chemical compound (where hydrogen has a valency of 1); e.g. sulphuric acid (H_2SO_4) has a valency of 2.

Vapour density is the **ratio** of the density of the vapour to the density of air. A dangerous example of this is where the vapours of flammable solvents are heavier than air and hence will flow into the lowest parts of a fume cupboard which could then cause a fire if there was a Bunsen burner in the near vicinity.

Verification of an **instrument** ensures that the instrument is correctly operated according to its stated manufacturer's operating instructions and specifications.

Vitreous meaning resembling of **glass** appearance or colour.

Voltammetry is a category of electroanalytical methods used in analytical chemistry where liquid substances must be oxidisable or reducible in the range where the solvent and electrode are electrochemically inert. It basically measures the **currents** generated in electrolytic solutions when known **voltages** are **applied**. However, it does not provide species identity but is often used to determine the amount of trace metals and toxins in water or other solutions.

Water, laboratory **grade** is generally classified into three types, namely I, II and III; or according to the ACS it is classified into four types, namely 1, 2, 3 and 4. The purest is type I, i.e. type 1 has maximum electrical conductivity of 0.06 µS/cm at 25 °C. This type is used for critical applications such as HPLC and IC. For preparing aqueous standards, use type II (type 2). The type for general laboratory work is type III having maximum electricity conductivity of 0.3 µS/cm at 25 °C. However, type 4 has maximum electricity conductivity of 5 µS/cm at 25 °C; this is the grade that should be used for rinsing out lab glassware before use.

Wilhelmy plate is a simple piece of apparatus that uses a thin plate to measure surface tension of liquids.

Quiz

Questions

1. Which of the following are homogeneous and which are heterogeneous?
   i. Vodka
   ii. Pure orange juice (unstrained)
   iii. Christmas cake
   iv. Table sugar

2. Why is it important to wear safety glasses in a laboratory?

3. A yoghurt sample has a pH of 4.56. Is it acidic or alkaline?

4. Which apparatus would you use for dispensing accurate volumes of liquid?
   i. 50 mL plastic measuring cylinder, or   ii. 50 mL glass burette

5. What is accuracy?

6. What is precision?

7. How many significant figures are there for the following pH readings taken during an experiment?
   i. 0.02   ii. 10.001   iii. 0.22   iv. 10.00

8. A 1 L salt solution contains 7 mg of NaCl. What is this result in?
   i. mg/L   ii. ppm

9. Does a solution of 0.1 N sulphuric acid have the same strength (concentration) as that of 0.1 mol/L sulphuric acid?

10. What is chemistry or chemical science?

11. What is the purpose of any laboratory?

12. You are required to sample a batch of 200 drums of orange juice; how would you proceed?

https://doi.org/10.1515/9783110721201-048

13. Fill in the details below from the bottle label below:

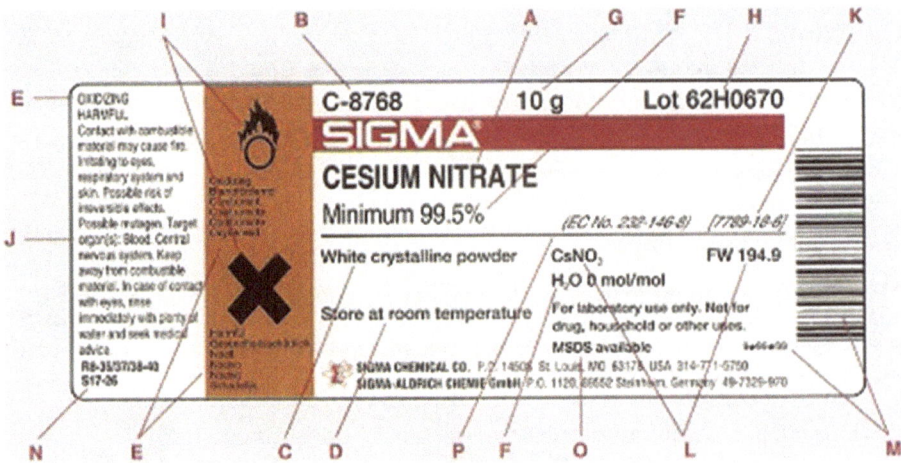

(Fig.quiz.13)

14. How would you measure the concentration of a sugar solution?

15. How would you clean burettes and pipettes?

16. You are required to take the temperature of a drum of frozen orange juice; how would you proceed?

17. A sample of refrigerated milk is required to be tested for its acidity level. How would you undertake the test?

18. A bottled spring water sample has a TDS by the oven-dried test method of 6,504 ppm. What is its expected electrical conductivity?

19. A suspension of fine sand in acidic mixture of 100 mL volume has to be filtered. What type and size of filter paper and funnel has to be used?

20. You are asked to measure the strength (i.e. concentration or dissolved solids) of, for example, sweetened apple juice. How would you proceed?'

21. What criteria are essential for accurate and precise viscosity measurements?

22. You are required to do a titration on acidity of orange juice; how are you going to perform the test?

23. What are the most dangerous hazards in a chemical laboratory?

24. How would you standardise a solution of approximately 0.1 N sodium hydroxide?

Answers

1. Which of the following are homogeneous and which are heterogeneous?
   i. Vodka – **homogeneous** – **one phase only**
   ii. Pure orange juice (unstrained) – **heterogeneous** – pulp present as well as liquid
   iii. Christmas cake – heterogeneous – various different types of raisins present (mixture)
   iv. Table sugar – homogeneous – all sugar crystals

2. Why is it important to wear safety glasses in a laboratory?
   To prevent splashes of acid or caustic or any other hazardous liquid from damaging your **eyes** (you have only one pair eyes – protect them and those of your fellow workers).

3. A yoghurt sample has a pH of 4.56. Is it acidic or alkaline?
   Acidic because pH is **below 7.**

4. Which apparatus would you use for dispensing accurate volumes of liquid?
   50 mL plastic measuring cylinder, or 50 mL glass burette?
   Burettes are for dispensing **accurate volumes** of liquid (e.g. during a titration) and **measuring cylinders** are for dispensing **approximate volumes** of liquid quickly into flasks, beakers, etc.

5. What is accuracy?
   How **close results** are to the true value or **most probable value**?

6. What is precision?
   How **close** results are to **each other**?

7. How many significant figures are there for the following pH readings taken on a weekly sample of effluent, for the month of January?
   i. One   ii. Five   iii. Two   iv. Four

8. A 1 L salt solution contains 7 mg of NaCl. What is this result in mg/L and in ppm?
   The question gives the answer: 7 milligrams in 1 litre; i.e. 7mg/L.
   PPM is parts per million, i.e. 1 part in a million (1,000,000) parts.
   Consider 1 gram of something to be approximately equivalent to 1 millilitre of volume (cf. 1 ml water weighs 1 gram, since its SG = 1).
   Now 7 milligrams are 0.007 grams and 1 litre are 1,000 millilitres. Therefore, 0.007 g/1,000 mL = 0.000007 = 7 in 1,000,000 = 7 ppm.

9. Does a solution of 0.1 N sulphuric acid have the same strength (concentration) as that of 0.1 mol/L sulphuric acid?

   No, because **sulphuric acid** has a **valency of 2** (i.e. **two hydrogen atoms**), therefore its normality is molarity/2. Thus, 0.1 N sulphuric acid solution is half as concentrated (not strong) as 0.1 mol/L sulphuric acid solution.

   That is 0.1 MH_2SO_4 = 0.2 NH_2SO_4

10. What is chemistry or chemical science?

    i. Chemistry may be considered to be the study and practice of the analytical breakdown and chemical properties and interactions of all parts of a substance (matter).

    ii. Chemical science is similar to analytical chemistry but also involves many other scientific disciplines such as biochemistry, microscopy, physics, statistics and applied mathematics, in order to fully understand all the properties, both physical and chemical, of the substance.

11. What is the purpose of any laboratory?

    A laboratory is a place where experiments are done and where the **data** therefrom are recorded, digitised, encrypted, interpreted and reported or published.

12. You are required to sample a batch of 200 drums of orange juice; how would you proceed (assume that the orange juice is homogeneous)?

    i. A possible answer to the above question is that the lab technician would select, at random, 10 drums out of the batch of 200 drums.

    ii. You would then withdraw approximately 250 mL sample (depending upon size of sampling apparatus) from each of these selected 10 drums; collecting these samples in a clean 5 L bucket with airtight lid (total volume collected should be about 2,500 mL). Mix well. Then transport as rapidly as possible to the laboratory for subsequent preparation of the analytical subsample.

    iii. Note that all sampling equipment including containers must be scrupulously clean (and sterile if any bacteriological tests are to be done).

    iv. Other possible answers could be:

    10% of the total number of drums (or containers); or as per an international (GATT) standard reference sampling procedure.

    Note that the more heterogeneous the type of product, the greater the number of samples that should be taken in order to get a more accurate representative primary sample of the batch/consignment.

13. Fill in the details below from the bottle label below:

    (Fig.QUIZ.13.CHEMICAL_LABEL)

    i. A = chemical name = caesium nitrate

    ii. B = manufacturer's catalogue or product number = C-8768

    iii. C = description of contents

    iv. D = storage conditions = store at room temperature

v.   E = hazard logos or symbols

vi. F = purity = 99.5%; for laboratory use only

vii. G = mass of contents or pack size = 10 g

viii. H = manufacturing lot number or batch number = Lot 62H0670

ix. I = GHS hazard symbols

x.   J = personal health risks and toxicity information

xi. K = Chemical Abstracts Service (CAS) identification number = 7789-18-6

xii. L = molecular formula = $CsNO_3$; molecular mass = 194.9

xiii. M = manufacturer's bar code tracking number

xiv. N = safety and risk phrases = R 8-36/37/38–40; S = 17–26

xv. O = Manufacturer's name and contact details = SIGMA Chemical Co

xvi. P = European Community identification number = EC No. 232-146-8

14. How would you measure the concentration of a sugar solution?

This can be done quickly and easily by using a **Brix hydrometer**, or a **refractometer** with degrees Brix scale. Published tables are available where RI values can be converted into Brix values.

Temperatures are critical here because hydrometer and refractometer readings are very much dependent upon temperature.

Brix, by **definition**, is the percentage of soluble sugars (soluble solids) present in a liquid.

Other methods are by chemical reactions with copper oxide solutions using the **reducing power of some sugars** (reduces cupric ions to cuprous oxide). An example is the common Lane Eynon Titration method with Fehling's Solutions A and B. Fehling's solutions also known as Soxhlet solutions.

15. How would you clean burettes and pipettes?

Here you must use whatever method as stated in the laboratory standard operating procedures (if available). Generally, common household dishwasher detergent is used.

This type of glassware must be scrupulously clean; otherwise, incorrect measurements will be made because of residual droplets of liquid adhering inside the burette or pipette after dispensing the required volume of liquid. Also, any detergent residues must also not be present as these will react (contaminate) with the liquid (reagent) being dispensed.

16. You are required to take the temperature of a drum of frozen orange juice; how would you proceed?

The substance, in this case, is frozen or very thick orange juice, thus a glass thermometer might break unless it is encased in a metal sheaf, also the temperature will be below 0 °C; therefore, you must use a thermometer that reads below 0 °C, and one that can penetrate frozen substances.

Thus, you should choose a digital thermometer with a pointed stainless steel probe covering the temperature range of, say −30 °C to ambient temperature or above. There are many different types of probes (thermocouples) available on the market for digital thermometers and for handheld infrared thermometers.

The thermometer (including digital thermometers) should also have been recently calibrated against a reference thermometer. These calibrations should be documented and logged according to your organisation's policies.

Note that when taking a temperature reading, the thermometer should be left in the sample for 1 to 2 min only, to get a correct equilibrated reading before the temperature of the sample changes.

17. A sample of refrigerated milk is required to be tested for its acidity level. How would you undertake the test?
   i. The lactic acid content can be tested by **titration** with a standardised 0.1 N solution of sodium hydroxide using phenolphthalein as indicator.
   ii. Alternatively, the **pH** level could be measured on a pH meter and the $[H^+]$ concentration is converted to mg/L of lactic acid using molecular mass factors. But milk has a buffering capacity, so in this case there is no direct correlation between "titratable acidity" and "pH".

18. A bottled spring water sample has a TDS by the oven-dried test method of 6,504 ppm. What is its expected electrical conductivity?
   i. TDS is the amount of total dissolved solids in a given volume of water (or liquid).
   ii. For an accurate calculation of the conductivity, one needs to know what cations (e.g. Ca^{2+} and Mg^{2+}) and anions (e.g. Cl^- and SO_4^-) are present in this sample of water. Each cation or anion has its own specific conductance (these values can be found in any chemistry reference book).
   iii. Check the result also by calculating a "chemical balance" of the water; also sum cation charges (+ve) must equal sum of anion charges (−ve).

19. A suspension of fine sand in acid of 100 mL volume has to be filtered. What type and size of filter paper and funnel has to be used?
   Acid-resistant paper, e.g. Whatman #541 of diameter 18.5 cm can be used to filter 100 mL suspension and filter funnel to be used will be 6 × 18.5 cm = 111 mm; that is, use a 110 mm diameter funnel.

20. You are asked to measure the strength (i.e. concentration or dissolved solids) of, for example, sweetened apple juice. How would you proceed?'
   i. Strength of fruit juices can be measured in degrees **Brix** (degrees Brix is the amount of dissolved sugar), by hydrometer or refractometer.
   ii. Alternatively, you could do a total solids **gravimetric** test (e.g. by oven drying).

21. What criteria are essential for accurate and precise viscosity measurements?
    i. Accurate **temperature** (in nearest 0.02 °C) and **timing** (in seconds) measurements; also, correct calibration of equipment.
    ii. And utilising the correct type of viscometer based on sample characteristics. Example: Brookfield or kinematic viscometer.

22. You are required to do a titration on acidity of orange juice; how are you going to perform the test?

    A possible answer to the above question is that you would first select a suitable **indicator**; in this case we will use phenolphthalein. We are expecting a high acidity (due to the presence of citric acid) so we will use a 0.1 N standardised solution of **sodium hydroxide**. These solutions can be bought already made and standardised, or you could prepare a solution from analytical-grade sodium hydroxide pellets and then standardised against standardised 0.1 N hydrochloric acid or with crystals of pure potassium hydrogen phthalate.

    We would then rinse a clean 25 mL burette with this solution. Then fill burette and adjust to the zero mark, making sure that there are no entrapped bubbles visible in the burette.

    Place a 10 mL aliquot of the orange juice, using a 10 mL pipette, into a conical flask, add a few drops indicator and titrate to the first faint pinkish colour; this is the end point of the titration.

    Note that if the orange colour is too dark to see the end point, then this titration is repeated using a pH electrode and meter in place of the indicator, i.e. a potentiometric titration is done.

    There are many automatic titrators available on the market.

    Note that this titration is a neutralisation reaction type, that is, a reaction between an acid (weak) and an alkali (strong) to form a salt (sodium citrate):

$$3NaOH + C_3H_4OH(COOH)_3 = C_3H_4OH(COONa)_3 + 3H_2O$$

The following calculation was used to obtain the result of an analysis (titration):

$$\frac{26.04 \text{ mL } (\pm 0.02 \text{ mL}) \times 0.0123 \text{ N } (\pm 0.0005 \text{ N}) \times 4.4236 \times 100}{2.683g(\pm 0.001 \text{ g})} = 52.808188266$$

(i.e. the display showed on a 12-digit electronic calculator)

Calculate the result to the correct number of figures and precision as follows:
    Now the % uncertainty of each factor or term is:

26.04 is (burette tolerance) $\pm \dfrac{0.02 \times 100}{26.04} = \pm 0.08\%$

And 0.0123 is (standardisation error) $\pm \dfrac{0.0005 \times 100}{0.0123} = \pm 4.1\%$

$$\text{And 2.683 is (mass meter tolerance)} \quad \pm \frac{0.001 \times 100}{2.683} = \pm 0.04\%$$

Note that the term 4.4236 is a "counting number" and has no uncertainty as it is derived from the stoichiometry of the titration; also, term 100 is percentage and thus has no uncertainty.

Therefore, the factor or term with the largest uncertainty (least precise) is 0.0123 N (three significant figures and has error of 4.1%); thus, the final result is 52.8 (three significant figures), and report this result as 52.8% ± 4% error.

The above method of calculating errors is only one of many other methods of calculation, e.g. the fish bone technique.

23. What are the most dangerous hazards in a chemical laboratory?
    i.   Glass cuts
    ii.   Acid/alkali burns
    iii. Chemical poisoning
    iv.   Mercury poisoning
    v.   Fires
    vi.   Electric shock
    vii. Static electricity
    viii. Explosions
    ix.   Radiation

24. How would you standardise a solution of approximately 0.1 N sodium hydroxide?

    By **titration** with either an accurately standardised solution of 0.1 N **hydrochloric acid** or a standard solution of exact known weight of **potassium hydrogen phthalate** crystals, using **phenolphthalein** as indicator.

Index

https://doi.org/10.1515/9783110721201-049